The Pocket Guide to
Critical Thinking

Fifth Edition

Richard L. Epstein

Illustrations by Alex Raffi

Advanced Reasoning Forum

Acknowledgements
I am grateful to Michael Rooney and Peter Adams for their help on this edition, and to the many others who have helped over the years on earlier editions.

For more information contact:
  Advanced Reasoning Forum
  P. O. Box 635
  Socorro, NM 87801 USA
  www.AdvancedReasoningForum.org

ISBN 978-1-938421-29-7 paperback

ISBN 978-1-938421-30-3 e-book

The Pocket Guide to Critical Thinking
Fifth Edition

Evaluating Risk, Making Decisions

Preface

Critical thinking is a set of skills that anyone can master. People who master these skills can see the consequences of what they and others say, they can formulate and communicate good arguments, and they can better make decisions.

Critical thinking is also the first step to writing well: first comes clear thinking, then comes clear writing.

The most important ideas and methods of critical thinking are presented here. But to reason well requires more than knowing definitions and rules and a few examples. It requires judgment and the ability to imagine possibilities. The practice you need for that can come from using these ideas every day while studying, watching television, reading the newspaper, browsing the Internet, working at your job, and talking to your friends and family.

Because your reasoning can be sharpened, you can understand more, you can avoid being duped. And, we hope, you will reason well with those you love and work with and need to convince, and you will make better decisions. But whether you will do so depends not just on method, not just on the tools of reasoning, but on your goals, your ends. And that depends on virtue.

Outline of the Book

Claims

We want to know truths from our reasoning. But to do that we need to know how to recognize whether a sentence could be true or false or is just nonsense, which is what we'll see in Chapter 1. In Chapter 2 we'll look at how to use definitions to clarify what we're talking about.

Arguments

In Chapter 3 we begin the study of reasoning by looking at arguments: attempts to convince someone that a claim is true in virtue of other claims being true. In Chapter 4 we'll see criteria for what counts as a good argument. And in Chapter 5 we'll see when we're justified in accepting a claim without an argument.

Most arguments we encounter aren't complete. That needn't mean they're bad, though, as we'll see in Chapter 6 when we set out criteria for repairing arguments. In Chapter 7 we'll see how to reply to objections with a counterargument. Sometimes, though, people try to get us to accept a claim by a fancy choice of words rather than reasoning, as we'll see in Chapter 8. In Chapter 9 we'll see that labeling certain kinds of arguments as fallacies can be a useful shortcut in evaluating arguments.

Special Kinds of Claims

Some kinds of claims require special skills to analyze in arguments. In Chapter 10 we'll look at claims that are made from other claims using "or," "not," and "if . . . then . . .". Especially important is learning how to formulate the contradictory of a claim. In Chapter 11 we'll look at claims about all or some part of a collection.

Claims that state not what is but what should be are crucial for reasoning about value and ethics, and we'll look at those in Chapter 12.

Numbers and Graphs

We use numbers to measure, summarize, and compare lots of information, and we'll see how to use those in our reasoning in Chapter 13. Graphs allow us to summarize many numerical claims, allowing for easier, visual comparisons, which we'll see in Chapter 14.

Reasoning about Experience

Comparisons are at the heart of our understanding of the world, and arguments that depend on those are called analogies, which we'll see how to evaluate in Chapter 15.

In Chapter 16 we'll see how to reason from our experience to arrive at true claims about a group from knowing about only a part of it.

We spend a lot of our time trying to figure out cause and effect in our lives, and in Chapter 17 we'll see how to do that well. In Chapter 18 we'll see how to analyze whether there is cause and effect by looking at studies of groups.

Reasoning in the Sciences

Reasoning in the sciences involves some distinctions and methods that supplement critical thinking skills. In Chapter 19 we'll see a method for looking for a cause. In Chapter 20 we'll see how scientists establish evidence with experiments. Chapter 21 is about explanations: what they are and how to evaluate them, which is as important in our daily lives as in science. Chapter 22 explains what models and theories are and how to judge them.

Risk and Making Decisions

A choice about what to do can be framed as an argument to convince yourself that a particular claim is true. To evaluate such reasoning, we need to be able to evaluate the risk as well as any benefit that might come from a choice of action, as we'll see in Chapters 23 and 24.

Writing Well

Knowing how to evaluate claims, arguments, cause and effect, and explanations can help us write better. We can judge our own work as we would another's, applying all we've learned here.

Cast of Characters

Claims

1 Claims

We want to know what's true. But first we have to recognize if a sentence even could be true or false.

> **Claims** A *claim* is a declarative sentence used in such a way that it is either true or false, but not both.

Example 1 Dogs are mammals.
> *Analysis* This is a claim.

Example 2 $2 + 2 = 5$
> *Analysis* This is a claim, a false claim.

Example 3 Dick is a student.
> *Analysis* This is a claim, even if we don't know if it's true.

Example 4 How can anyone be so dumb to think cats can reason?
> *Analysis* This is not a claim. Questions are not claims.

Example 5 Never use gasoline to clean a hot stove.
> *Analysis* Instructions and commands are not claims.

Example 6 I wish I could get a job.
> *Analysis* Whether this is a claim depends on how it's used. If Maria, who's been trying to get a job for three weeks says this to herself, it's not a claim—we don't say that a wish is true or false. But if Dick's parents are berating him for not getting a job, he might say, "It's not that I'm not trying. I wish I could get a job." Since he could be lying, in that context it's a claim.

Vagueness

Often what people say is *too vague* to take as a claim: there's no single obvious way to understand the words. Vagueness can create worthless disagreements and mislead.

Example 7 People who are disabled are just as good as people who aren't.
> *Analysis* Lots of people take this to be true and important, but what does it mean? A deaf person is not as good as a hearing person at letting people know a smoke alarm is going off. This is too vague for us to agree that it's true or false.

Example 8 "Susan Shank, J.D., has joined Zia Trust Inc. as Senior Trust Officer. Shank has 20 years' experience in the financial services industry including 13 years' experience as a trust officer and seven years' experience as a wealth strategist."

Albuquerque Journal, 4/29/10 and the Zia Trust website

Analysis "Wealth strategist" looks very impressive. But when asked what it meant, Ms. Shank said, "It can have many meanings, whatever the person wants it to mean." This is vagueness used to convince you she's doing something important.

Still, everything we say is somewhat vague. After all, no two people have identical perceptions, and since the way we understand words depends on our experience, we all understand words a little differently. So it isn't whether a sentence is vague, but whether it's too vague, given the context, for us to be justified in saying it's a claim. It's a mistake, a ***drawing the line fallacy***, to argue that if you can't make the difference precise, there's no difference. In a large auditorium lit by a single candle at one end, there's no place where we can say it stops being light and starts being dark. But that doesn't mean there's no difference between light and dark.

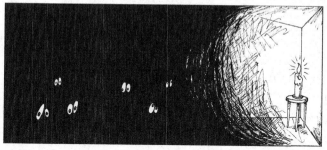

Example 9 Tom: My English composition professor showed up late for class today.

Zoe: What do you mean by late? How do you determine when she showed up? When she walked through the door? When her nose crossed the threshold?

Analysis Zoe is asking for more precision than is needed. In ordinary talk what Tom said is clear enough to be a claim.

Example 10 If a suspect who is totally uncooperative is hit once by a policeman, that's not unnecessary force. Nor twice, if he's resisting. Possibly three times. If he's still resisting, shouldn't the policeman have the right to hit him again? It would be dangerous not to allow that. So, you can't say exactly how many times a policeman has to hit a suspect before it's unnecessary force. So the policeman did not use unnecessary force.

Analysis This argument convinced a jury to acquit the policemen who beat up Rodney King in Los Angeles in the 1990s. But it's just an example of the drawing the line fallacy.

Example 11 Zoe: Those psychiatrists can't agree whether Wanda is crazy or not. One says she's clinically obsessive, and the other says she just likes to eat a lot. This psychiatry business is bunk.

Analysis Just because there are borderline cases doesn't mean there isn't a clear difference between people who are really insane and those who aren't.

Subjective claims

It's useful to distinguish between claims that are about the world outside us and those about thinking, believing, and feeling.

> **Subjective and objective claims** A claim is *subjective* if whether it's true or whether it's false depends on what someone, or something, or some group thinks, believes, or feels. A claim that's not subjective is *objective*.

Example 12 All ravens are black.
  Analysis This is an objective claim.

Example 13 Dick: My dog Spot is hungry.
  Analysis This is a subjective claim.

Example 14 Suzy: It's cold outside.
  Analysis This is too vague to be an objective claim. But if Suzy means just that it seems cold to her, it's a subjective claim. A sentence that's too vague to be an objective claim might be perfectly all right as a subjective one if that's what the speaker intended. After all, we don't have very precise ways to describe our feelings.

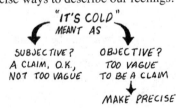

Example 15 Lee: Calculus I is a really hard course.

Analysis What standard is Lee using for classifying a course as really hard? If he means that Calculus I is difficult for him, then the claim is subjective. If Lee has in mind that about 40% of students fail Calculus I, which is twice as many as in any other course, then the claim is objective. Or Lee might have no criteria in mind, in which case what he's said is too vague to be taken as a claim. *If it's not clear what criteria are being invoked, then the sentence is too vague to be classified as a claim.*

Example 16 Lee: I felt sick yesterday, and that's why I didn't hand in my work.

Analysis Lee didn't feel sick yesterday—he left his critical thinking writing assignment to the last minute and couldn't finish it before class. This is a subjective claim, but a false one.

Example 17 Wanda weighs 215 pounds.

Analysis This is an objective claim. Registering a number on a scale is an objective criterion.

Example 18 Nurse: Dr. E, tell me on a scale of 1 to 10 how much your back hurts.

Dr. E: It's about a 7.

Analysis This is a scale, but one that only Dr. E knows. Dr. E's claim is subjective.

Example 19 Dick: Wanda is fat.

Analysis This is a subjective claim. Whether it's true depends on Dick's feeling about what is fat. But what if Wanda is so obese that everyone would consider her fat? It's still subjective, but we ought to note that agreement. A subjective claim is ***intersubjective*** if (almost) everyone agrees that it's true or (almost) everyone agrees that it's false.

Example 20 God exists.

Analysis Often people think that a lot of disagreement about whether a claim is true means the claim is subjective. But that's a confusion, the ***subjectivist fallacy***. Whatever we mean by "God" it's supposed to be something that exists independently of people. So the example is objective: whether it's true or false doesn't depend on what anyone thinks or feels. "God exists" ≠ "I believe that God exists."

Example 21 There are an even number of stars in the sky.

   Analysis This claim is objective, but no one knows how to find out whether it's true or false, and it's not likely we'll ever know.

Example 22 There is enough oil available for extraction by current means to fulfill the world's needs for the next 43 years at the current rate of use.

   Analysis This is objective. People disagree about it because there's not enough evidence one way or the other.

Example 23 Zoe (to Dick): Tom loves Suzy.

   Dick: I don't think so.

   Analysis Dick and Zoe disagree about whether this subjective claim is true, but it's not for lack of evidence. There's plenty; the problem is how to interpret it.

Whether a claim is objective or subjective does not depend on:

- How many people believe it.
- Whether it's true or false.
- Whether anyone can know whether it's true or whether it's false.

To evaluate any claim we have to use our judgment. When we reckon that too much judgment is needed, it's usually because the sentence is too vague to be a claim.

Confusing whether a claim is objective or subjective can lead to pointless disagreements.

Example 24

   Analysis Dick and Zoe are treating a subjective claim as objective. There's no sense in arguing about taste.

Example 25 Lee: I deserve a higher mark in this course.

Dr. E: No you don't. Here's the record of your exams and papers.
Summing them all up, you earned a C.

Lee: That's just your opinion.

Analysis Lee is treating an objective claim, "I deserve a higher mark," as if it were subjective. But if it really were subjective, there would be no point in arguing about it with Dr. E any more than arguing about whether Dr. E feels hungry.

Often it's reasonable to question whether a claim is really objective. But sometimes it's just a confusion. All too often people insist that a claim is subjective ("That's just your opinion") when they are unwilling to examine their beliefs or engage in dialogue.

2 Definitions

To reason well we need to understand the words that we and others use.

> **Definitions** A *definition* is an explanation or stipulation
> of how to use a word or phrase.

A definition is not a claim. A definition is not true or false,
but good or bad, right or wrong. Definitions tell us what we're
talking about.

Example 1 "Exogenous" means "developing from without."
  Analysis This is a definition, not a claim. It's an explanation
of how to use the word "exogenous."

Example 2 Puce is the color of a flea, purple-brown or brownish-
purple.
  Analysis This is a definition, not a claim.

Example 3 Lee: Maria's so rich, she can afford to buy you dinner.
  Tom: What do you mean by "rich"?
  Lee: She's got a Mercedes.
  Analysis This is not a definition—or it's a very bad one. Some
people who have a Mercedes aren't rich, and some people who are rich
don't own a Mercedes. That Maria has a Mercedes might be some
evidence that she's rich.

Example 4 "Fasting and very low calorie diets (diets below 500
calories) cause a loss of nitrogen and potassium in the body, a loss
which is believed to trigger a mechanism in the body that causes us
to hold on to our fat stores and to turn to muscle protein for energy
instead."   *Jane Fonda's New Workout and Weight Loss Program*
  Analysis Definitions aren't always labeled but are often made in
passing, as with this good definition of "very low calorie diet."

What's a good definition?

Example 5 "Intuition is perception via the unconscious."
                                    Carl G. Jung
  Analysis This is a definition, but a bad one. The words doing the
defining are no clearer than what's being defined.

Example 6 A car is a vehicle with a motor that can carry people.

Analysis This is a bad definition because it's **too broad**: it covers cases that it shouldn't. In this case, a golf cart would be classified as a car. So we can't use the words doing the defining in place of the word being defined.

Example 7 Dogs are mammals.

Analysis This is not a definition but a claim. We can't use "mammal" in place of "dog" in our reasoning.

Example 8 Dogs are domesticated canines that obey humans.

Analysis This is a bad definition because it's **too narrow**: it doesn't cover cases it should, like feral dogs in India.

Good definition For a definition to be good:

- The words doing the defining are clear and better understood than the word or phrase being defined.
- It would be correct to use the words doing the defining in place of the word or phrase being defined. That is, the definition is neither too broad nor too narrow.

Example 9 Abortion is the murder of unborn children.

Analysis Here what should be debated—whether abortion is murder—is being assumed as if it were a definition. A **persuasive definition** is a contentious claim masquerading as a definition.

Example 10 A feminist is someone who thinks that women are better than men.

Analysis This is a persuasive definition.

"If you call a tail a leg, how many legs has a dog? Five?
No, calling a tail a leg don't *make* it a leg."

                                    Attributed to Abraham Lincoln

Example 11 "Absurdity: A statement of belief manifestly inconsistent with one's own opinion."   Ambrose Bierce, *The Devil's Dictionary*

Analysis Whether you classify this as persuasive depends on how much faith you have in people.

To make a good definition we need to look for examples where the definition does or does not apply to make sure it's not too broad or too narrow.

Example 12   Suppose we want to define "school cafeteria." That's something a lawmaker might need in order to write a law to disburse funds for a food program. As a first go, we might try "A place in a school where students eat." But that's too broad, since that would include a room with no food service where students can take their meals. So we could try "A place in a school where students can buy a meal." But that's also too broad, since it would include a room where students could buy a sandwich from a vending machine. How about "A room in a school where students can buy a hot meal that is served on a tray"? But if there's a fast-food restaurant like Burger King at the school, that would qualify. So it looks like we need "A room in a school where students can buy a hot meal that is served on a tray, and the school is responsible for the preparation and selling of the food." This looks better, though if adopted as a definition in a law it might keep schools that want money from the legislature from contracting out the preparation of their food. Whether the definition is too narrow will depend on how the lawmakers intend the money to be spent.

Steps in making a good definition
- Show the need for a definition.
- State the definition.
- Make sure the words make sense and are clear.
- Give examples where the definition applies.
- Give examples where the definition does not apply.
- If necessary, contrast it with other likely definitions.
- If necessary, revise it.

Arguments

3 Arguments

> **Arguments** An *argument* is an attempt to convince someone,
> possibly yourself, that a particular claim, called the **conclusion**,
> is true. The rest of the argument is one or more other claims,
> called the **premises**, which are given as the reasons for believing
> that the conclusion is true.

Example 1 Critical thinking is the most important subject you'll ever
study. It will help you reason better, it will help you get a job, and it
will help you make better decisions.

    Analysis This is an argument. The conclusion is "Critical think-
ing is the most important subject you'll ever study." The premises are
"Critical thinking will help you reason better," "Critical thinking will
help you get a job," and "Critical thinking will help you make better
decisions."

Example 2 Suzy (to Tom): You can tell that economics graduates are
smart. They get high-paying jobs, and they always dress well.

    Analysis This is an attempt by Suzy to convince Tom or maybe
just herself that "Economics graduates are smart" is true. Its premises
are "They get high paying jobs" and "They always dress well."

Example 3

    Analysis Dick is making an argument, trying to convince the police
officer that the following claim is true: "The accident was not my fault"
(reworded a bit). He uses two premises: "She hit me from the rear" and
"Anytime you get rear-ended it's not your fault."

Example 4 (From a label on a medication) Follow the directions for using this medicine provided by your doctor. This medicine may be taken on an empty stomach or with food. Store this medicine at room temperature, away from heat and light.

Analysis This is not an argument. Instructions or commands are not an attempt to convince anyone that some claim is true.

Example 5

HOW COME YOU DON'T
CALL ME? WHAT'S
WRONG? YOU DON'T
LOVE YOUR MOTHER?
WHERE DID I GO
WRONG?...

Analysis Zoe's mother is attempting to convince her, but not of the truth of a claim. So there's no argument.

Example 6 If it's OK to buy white mice to feed a pet boa constrictor, why isn't it OK to experiment on rats?

Analysis This isn't an argument, since questions aren't claims. We might construe it as an attempt to convince, taking the question as rhetorical. But before we go putting words in people's mouths, we should have a clearer idea of when we're justified in re-interpreting what they say as an argument.

Example 7 Dick: You shouldn't dock your dog's tail because it will hurt her, it'll make her insecure, and she won't be able to express her feelings.

Analysis This is an argument. The word "because" clues us to that by setting off the premises.

An ***indicator word*** is a word or phrase added to a claim to tell us the role of the claim in an argument or what the speaker thinks about the claim or argument. Here are some examples.

Conclusion indicators hence; therefore; so; thus; consequently; we can then show that; it follows that

Premise indicators since; because; for; in as much as; due to; given that; suppose that; it follows from; on account of

Indicators of speaker's belief probably; certainly; most likely; I think

Example 8   (1) Maria knows a lot about nursing, all the details about taking care of patients. Probably she's worked in a hospital.

(2) It seems to me that Maria knows a lot about nursing, all the details about taking care of patients. So I'm sure she's worked in a hospital.

Analysis   These are the same argument. The indicator words show us what the speaker thinks about the argument, but they aren't part of the argument.

4 What Is a Good Argument?

A good argument should give us good reason to believe its conclusion. But what's "good reason"?

The premises are plausible

If we don't have good reason to believe the premises, they can't give us good reason to believe the conclusion. From a false premise we can as easily derive a false conclusion as a true one.

Example 1 The Prime Minister of England is a dog. All dogs have fur. So the Prime Minister of England has fur.
   Analysis This has a false premise and its conclusion is false.

Example 2 The Prime Minister of England is a dog. All dogs have a liver. So the Prime Minister of England has a liver.
   Analysis This has a false premise, yet its conclusion is true.

> ***Plausible claims*** A claim is *plausible* if we have good reason to believe it's true. It is less plausible the less reason we have to believe it's true. An implausible claim is also called *dubious*.

An argument is no better than its least plausible premise.

Example 3 Suzy: Dr. E is mean.
   Wanda: Why do you say that?
   Suzy: Because he's not nice.
   Analysis Suzy's given no reason for Wanda to believe that Dr. E is mean because "He's not nice" is not more plausible than that.

Example 4 Maria (to Dick): Every dog has a soul. So you should treat dogs humanely.
   Analysis The conclusion is plausible to Dick. He reckons the premise is plausible, too. But it's not more plausible to him than the conclusion. So Maria's argument gives him no more reason to believe the conclusion than he had before she spoke.

> ***Begging the question*** An argument *begs the question* if even one of its premises is not more plausible than its conclusion.

The conclusion follows from the premises

Even if the premises are plausible and more plausible than the
conclusion, they won't give good reason to believe the conclusion
unless the conclusion follows from them.

Example 5 Dollar bills are printed using green ink. Therefore, U.S.
currency is easy to counterfeit.

Analysis Our reaction to the single true premise here is "So?"
When you ask "So?" you're asking why the conclusion follows from
the premises.

Example 6 This book teaches how to reason. So this book costs less
than $50.

Analysis The premise of this is clearly true. But it gives no
reason to believe the conclusion. This book is so valuable to students
that it might be priced at $100.

Example 7 Dick: All students here pay tuition. Harry is a student.
So Harry pays tuition.

Analysis If the premises are true, then there's no way for the
conclusion to be false—it's just impossible. The conclusion certainly
follows from these premises.

Example 8 Maria's hair is naturally black. Today Maria's hair is red.
So Maria dyed her hair.

Analysis It's not impossible for the premises to be true and
conclusion here to be false: Maria might be taking a new medication
that has a strong effect, or she might have been too close to her car
when they were painting it. But those are very unlikely. Unless
someone can come up with a more likely way the premises could be
true and conclusion false, we can say the conclusion follows here.

Valid arguments An argument is *valid* if it is impossible for
the premises to be true and conclusion false at the same time;
otherwise it is *invalid*.

Strong and weak arguments An argument is *strong* if there
are ways for the premises to be true and conclusion false at the
same time, but they're all unlikely. An invalid argument that is
not strong is *weak*.

To say that the conclusion of an argument *follows from* the premises means that the argument is valid or strong.

Example 9 All dogs bark. Ralph is a dog. Therefore, Ralph barks.

Analysis This is a valid argument: there's no way for the premises to be true and conclusion false at the same time. But the argument is bad because the first premise is false: Basenjis can't bark, and some dogs have had their vocal cords cut. Whether an argument is valid or strong depends on the relation between the premises and conclusion, not on whether the premises happen to be true. *Valid ≠ Good*.

Example 10 Good teachers give fair exams. Dr. E gives fair exams. So Dr. E is a good teacher.

Analysis This is a weak argument. Dr. E might bore his students to tears and just copy good exams from the instructor's manual, or he might get good exams from another teacher. There are lots of ways the premises could be true and conclusion false that are not unlikely.

Either an argument is valid or it isn't. But *the strength of an invalid argument is a matter of degree* according to how likely it is that the premises could be true and conclusion false.

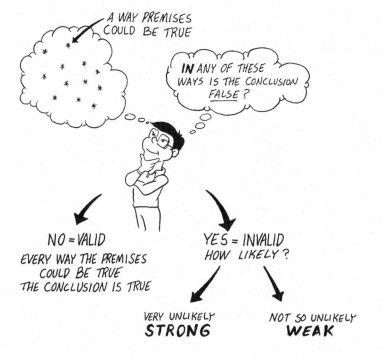

Why should we worry about whether the conclusion follows from the premises if we don't know that the premises are true? Consider what happens when Lee goes to apply for a student loan. He fills out all the forms and gives them to the loan officer at the bank. She reads his answers. At that point she might tell him that he doesn't qualify. That is, even though she doesn't know if the claims he made about his income and assets are true, she can see that even if they are true, Lee won't qualify for a loan. On the other hand, she might tell him that he'll qualify—if those claims are true. Then she'll have to make phone calls, check credit references, and so on to find out if what he claimed is true. It's the same with arguments. Sometimes it's easier to evaluate first whether the conclusion follows from the premises—whether an argument is valid or strong—in order to find out whether we should bother to investigate whether the premises are true.

How do we show an argument is *not* valid or strong? We give an example, a description of how the world might be where the premises would be true and conclusion false. *To reason well we must use our imagination.*

We've now seen what is needed for an argument to be good.

> *Tests for an argument to be good*
> - The premises are plausible.
> - The premises are more plausible than the conclusion.
> - The argument is valid or strong.

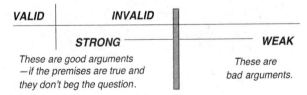

VALID	INVALID	
	STRONG ————	———— **WEAK**
	These are good arguments —if the premises are true and they don't beg the question.	These are bad arguments.

- Every good argument is valid or strong.
- Not every valid or strong argument is good.
  (It could have a dubious premise or beg the question.)
- Only invalid arguments are classified from strong to weak.
- Every weak argument is bad.
- If the conclusion of a valid argument is false, one of its premises must be false.

Examples

Example 11 Dick: All parakeets I've ever seen or heard or read about in bird books and the encyclopedia are under 2 feet tall. So the parakeets for sale at the mall are under 2 feet tall.

Analysis Surveying all the ways the premise could be true, we think that yes, a new supergrow bird food could have been formulated and the parakeets at the local mall are really 2 feet tall — we just haven't heard about it. Or a giant parakeet from the Amazon forest could have been discovered and brought here. Or a UFO might have abducted a parakeet by mistake, hit it with growing rays, and now it's really big. So the argument is not valid. But all these ways the premise could be true and the conclusion false are so unlikely that we would have very good reason to believe the conclusion if the premise is true, though the conclusion might be false. This is a strong argument.

Example 12 All parakeets are under 2 feet tall. Therefore, the parakeets for sale at the mall are under 2 feet tall.

Analysis This argument is valid, but it's not better than the previous one. The only way we have of showing that the premise is plausible is with the premise we used in the previous argument. What good is validity if the premise is dubious? *When reasoning from experience, a strong argument is often better than a valid one.*

Example 13 Dr. E is a philosophy professor. All philosophy professors prefer dogs to cats. So Dr. E prefers dogs to cats.

Analysis This is a valid argument: it's impossible for the premises to be true and the conclusion false. And the conclusion is true. But it's bad because the second premise is false.

Example 14 Dick is a bachelor. So Dick was never married.

Analysis There's a likely way the premise could be true and the conclusion false: Dick could be divorced. This is a weak argument and hence bad.

Example 15 Whenever Spot barks, there's a cat outside. Since he's barking now, there must be a cat outside.

Analysis This is a bad argument: Spot might be barking at the garbageman. That doesn't show the argument is weak — it shows that the first premise is false. This is a valid argument.

Example 16 Maria (to her supervisor): I was told that if I put in 80 hours of overtime and have a perfect attendance record for two months, then I'll get a bonus. I've put in over 100 hours of overtime and I haven't missed a day for the last 11 weeks. So I'm entitled to a bonus.

Analysis The argument is valid. But we don't know if it's good because we don't know if the premises are true.

Example 17 Prosecuting attorney: The defendant intended to kill Louise. He bought a gun three days before he shot her. He practiced shooting at a target that had her name written across it. He staked out her home for two nights. He shot her twice.

Analysis This is a strong argument. It's good, *if* the premises are plausible.

Example 18 All the dockworkers at Boa Vista docks belong to the union. Luis has been working at Boa Vista docks for two years. So probably Luis belongs to the union, too.

Analysis "Probably" is an indicator word about the speaker's belief. The conclusion here is "Luis belongs to the union," and the argument is valid—regardless of what the speaker thought.

Whether an argument is valid or strong does *not* depend on:

— whether the premises are true.
— whether we know the premises are true.
— whether the person making the argument thinks the argument is valid or strong.

A good argument gives us good reason to believe the conclusion. But *a bad argument tells us nothing about whether the conclusion is true or false.* If we encounter a bad argument, we have no more reason to believe or disbelieve the conclusion than we had before. *A bad argument does* not *show that the conclusion is false or even doubtful.*

5 Evaluating Premises

An argument gives us good reason to believe a claim if we have good reason to believe its premises. But what are good reasons to believe the premises? We can't expect an argument for every claim or we'd never get started. We have to take some claims without argument, and we need criteria for when it's OK to do that. But keep in mind:

not believe $\neq$ believe is false

lack of evidence $\neq$ evidence it's false

We might have no good evidence that a particular claim is true or that it's false, in which case we should suspend judgment.

Three choices we can make about whether to believe a claim
- Accept the claim as true.
- Reject the claim as false.
- Suspend judgment.

Criteria for accepting an unsupported claim

Our most reliable source of information about the world is our own experience.

We need to trust our own experience because that's the best we have. Everything else is secondhand. Should you trust your buddy, your spouse, your priest, your professor, the President when what they say contradicts what you know from your own experience? That way lies demagoguery, religious intolerance, and worse. Too often leaders have manipulated the populace: All Muslims want to overthrow the U.S.? But what about my Muslim neighbor who's on the city council? You have to forget your own experience to believe the Big Lie. They repeat it over and over and over again until you begin to believe it, even when your own experience says it isn't so.

Oh, we get the idea. Don't trust the politicians. No. It's a lot closer to home than that. Every rumor, all the gossip you hear, compare it to what *you* know about the person or situation. Don't repeat it. Be thoughtful, not part of the humming crowd.

"Who are you going to believe, me or your own eyes?"

Chico Marx

But we shouldn't always trust our own experience.

Example 1   As Sgt. Carlson of the Las Vegas Police Department says, "Eyewitnesses are terrible. You get a gun stuck in your face and you can't remember anything." The police do line-ups, putting a suspect to be identified by a witness among other people who look a bit similar. The police have to be careful not to say anything that may influence the witness, because memory is malleable.

Example 2   You tell the officer that the car ahead didn't put on its turn signal.

Analysis   You think that's so, but with the rain and distractions you might have missed it. The state of the world around us can affect our observations and make our personal experience unreliable.

Example 3   You go to the circus and see a magician cut a lady in half. You saw it, so it has to be true.

Analysis   You don't believe it, and rightly so because it contradicts too much else you know about the world.

Example 4   Day after day we see the sun rise in the east and set in the west. So clearly the sun revolves around the Earth.

Analysis   We don't accept our own experience because there's a long story, a theory of how the Earth turns and revolves around the sun. A convincing argument has been made for us to reject our own experience, and that argument builds on other experiences of ours.

- We accept a claim if we know it is true from experience.
- We reject a claim if we know it is false from experience.

    Exceptions   We have good reason to doubt our memory.
                        The claim contradicts what we know from other
                        experiences and there's a good argument against it.

Example 5   Tom was asked to describe what he sees in the picture.

He wrote: "The guy is in the room and he spots a purse on the table. He looks around pretty shiftily and thinks that he can get away with taking the purse. So he grabs it and goes."

Analysis How does Tom know that the guy thinks he can get away with it? How does Tom know the guy grabbed the purse? Tom didn't see that. Perhaps the purse belongs to the guy's girlfriend and he was looking around for her, and then he took it to her. *Personal experience means what we perceive—not what we deduce from that.*

Example 6 Wanda: Chinese guys are really smart. There are five of them in my calculus class and they're all getting an A.

Analysis It's not Wanda's personal experience that all Chinese guys are really smart, but a deduction she's made from knowing five of them.

We can accept a claim made by someone we know and trust who knows about this kind of claim.

Example 7 Zoe tells Dick to stay away from the area around South Third and Westermeyer Avenue. She's seen people doing drugs there and knows two people who were mugged at that corner.

Analysis Dick has good reason to believe Zoe's claims.

Example 8 Dick's mother tells him that he should major in business so he can get ahead in life.

Analysis Should Dick believe her? She can tell him about her friends' children. But what really are the chances of getting a good job with a degree in business? It would be better to check at the local colleges where they keep records on what jobs graduates get. Dick shouldn't reject her claim; he should suspend judgment until he gets more information.

We can accept a claim made by a reputable authority we can trust as an expert on this kind of claim and who has no motive to mislead.

Example 9 Compare:

- The Surgeon General announces that smoking is bad for your health.
- The doctor hired by the tobacco company says there's no proof that smoking is addictive or causes lung cancer.
- The new Surgeon General says that marijuana should be legal.

Analysis The Surgeon General is a reputable physician with expertise in public health. She's in a position to survey the research on the subject. We have no reason to suspect her motives. So it's reasonable to believe her.

But is the doctor hired by the tobacco company an expert on smoking-related diseases, or an allergist, or a pediatrician? He has motive to mislead. There's no reason to accept his claim.

Nor is there any reason to accept what the Surgeon General says about what should be law. Though she's an authority on health, she's not an expert on law and society.

We can accept a claim put forward in a reputable journal or reference source.

Example 10   The New England Journal of Medicine is regularly quoted in newspapers, and for good reason. Articles in it are reviewed by experts who are asked to evaluate whether the research was done to scientific standards. We have less reason to trust *The National Geographic* because it pays for its own research in order to sell the magazine. What about *Scientific American*? Are the articles there peer-reviewed or commissioned? And anyone can incorporate as the "American Institute for Global Warming Analysis" or any other title you like. A name is not enough to go by.

We can accept a claim from some usually reliable media source that has no obvious motive to mislead, if the person being quoted is named.

It's up to you to decide from experience whether a source is reliable. Don't trust a news report that makes that decision for you by quoting unnamed "usually reliable sources." They're not even as reliable as the person who's quoting them, and anyway they've covered themselves by saying "usually." If there's reliable information there, the reporter should be able to back it up with documents or quotes. Otherwise, it's just rumor, often planted to sway opinion. *There's never good reason to believe a claim from an unnamed source.*

Look also for bias in the media source because of its advertisers. Ask yourself, "Who will benefit from my believing this?"

There are no absolute rules for when to accept, when to reject, and when to suspend judgment about a claim. It's a skill, weighing up these criteria in order of importance.

Criteria for judging unsupported claims

Accept   The claim is known to be true from personal experience. *Exceptions* Our memory is not good; there's a good argument against what we thought was our experience; it's not our experience but what we've concluded from it.

Reject   The claim is known to be false from personal experience.

Reject   The claim contradicts other claims we know to be true.

Accept   The claim is made by someone we know and trust who knows about this kind of claim.

Accept   The claim is made by a reputable authority we can trust as an expert about this kind of claim and who has no motive to mislead.

Accept   The claim is made in a reputable journal or reference source.

Accept   The claim is from some media source that's usually reliable and has no obvious motive to mislead, if the person being quoted is named.

We don't have criteria for when to suspend judgment. That's the default attitude we adopt whenever we don't have good reason to accept or reject a claim.

Above all, personal experience is your best guide. Don't trust others more than yourself about what you know best.

Example 11   A Nevada couple letting their SUV's navigation system guide them through the high desert of eastern Oregon got stuck in snow for three days when their GPS unit sent them down a remote forest road.
Albuquerque Journal, 12/29/09

Analysis How far down a remote snow-packed forest road do *you* have to go before you trust your own senses over your GPS unit?

Advertising

Advertisements are meant to convince you of the (often unstated) claim that you should buy the product, or frequent the establishment, or use the service. Sometimes the claims are accurate, sometimes they're not. There's nothing special about them, though. They should be judged by the criteria we've already considered.

If you think there should be more stringent criteria for evaluating ads, you're not judging other claims carefully enough.

Example 12 "Gold is the only asset that's not somebody else's liability."   Radio advertisement, Spring 2010

Analysis That's false: when you've paid off your car it's not someone else's liability.

Example 13 "We're Credit Card Relief . . . We've been helping people like you for more than a decade. We're an attorney-driven program."   Radio advertisement, Spring 2010

Analysis This isn't true or false: "attorney-driven program" is meaningless, though it sounds impressive.

Example 14 "Wendy's. Our beef is fresh. Never frozen."
                    Billboard in Albuquerque, NM

Analysis So? Are we supposed to believe that fresh is better?

The Internet

What reason do you have to believe something you read on the Internet? Next time you're ready, mouth agape, to swallow what's up there on the screen, imagine your friend saying, "No, really, you believed *that*?" Don't check your brain at the door when you go online.

Example 15 "Colonial records refer to small, nearly hairless dogs at the beginning of the 19th century, one of which claims 16th-century Conquistadores found them plentiful in the region later known as Chihuahua.*

* Pedro Baptista Pino y Juan Lopez Cancelada, *Exposición sucinta y sencilla de la Provincia del Nuevo México y otros escritos*. Ed. Jesus Paniagua Perez. Valladolid: Junta de Castilla / León: Universidad de León, 2007, p. 244: "even in the desert the tiny dogs could be found, hunting rats, mice, and lizards." The footnote that follows alludes to starving Conquistadores reportedly hunting and stewing the dogs (Universidad Veracruzana, Arquivo Viejo, XXVI.2711)."   "Chihuahua dog," *Wikipedia*, February 2012

Analysis You believe what you read on Wikipedia, and quote it, too. Only this is pure fantasy written by Michael Rooney and me. There is such a book but there's no such quote, and we made up the reference to Arquivo Viejo. What makes you think that any other entry in Wikipedia is more reliable?

Wikipedia is not an encyclopedia. Entries aren't signed, so you can't evaluate the expertise of the author. Lots of ignorant people correcting each other does not result in a reliable source. At best, Wikipedia entries are useful to stimulate our imaginations and provide references we can consult.*

Common mistakes in evaluating claims

Example 16 Tom: All CEOs of computer software companies are
                            rich. Bill Gates is a CEO of a computer software company.
                            So Bill Gates is rich.

          Suzy: Gee, that's valid, just like Dr. E said. And Bill Gates is sure
                            rich. So I guess all CEOs of computer software companies
                            are rich.

Analysis Suzy's arguing backwards. An argument is supposed to convince that its conclusion is true, not that its premises are true. There are lots of CEOs of small software companies who are working hard just to make a living.

Arguing backwards The fallacy of *arguing backwards* is to reason that because an argument is clearly valid or strong and the conclusion is true, the premises must be true, too.

Sometimes we have good reason to believe a claim because it's put forward by an authority. But it's a mistake to accept a claim when the person isn't an authority on the subject or has a motive to mislead: that's a **bad appeal to authority**.

Example 17 Zoe: What do you think of the President's new science
                            funding plan?

          Tom: It's awful. It'll cut back funding on military research. They
                            said so on Fox News.

Analysis Not everything you hear on Fox News is true.

Though it's OK to suspend judgment on a claim if you don't

---

* See <http://chronicle.com/article/The-Undue-Weight-of-Truth-on/130704/> for an excellent critique of how ignorance can trump expertise in Wikipedia. And see <http://www.newyorker.com/archive/2006/07/31/060731fa_fact> for unmasking an editor of Wikipedia.

consider the person who's making it to be a reputable authority, it's never right to say a claim is false because of who said it. That's *mistaking the person (or group) for the claim*.

Example 18   Tom: I don't believe the new global warming accord will help the environment. That's just another lie our President said.

Dick: Come on, it's not false just cause he said it. Politicians don't lie all the time.

Analysis Tom is mistaking the person for the claim. There's no shortcut for thinking about a claim in order to evaluate whether to accept it.

Example 19   Tom: There's no water shortage here in New Mexico. That's just one of those things environmentalists say.

Analysis Here Tom is mistaking the group for the claim.

An *appeal to common belief* is to accept a claim as true because a lot of other people believe it. Typically that's a bad appeal to authority.

Example 20   Lee: All the guys at my work say that Consolidated Computers is a great investment. So I'm going to buy 500 shares—they can't all be wrong.

Analysis Lee is making an appeal to common belief, which is just a bad appeal to authority.

The standard fare of conspiracy theorists is to think that because it's possible, it's true. Just because it could happen, and you don't trust the folks who would benefit if you don't believe it, doesn't make it true. *Possibility isn't plausibility*.

Example 21   Tom: Terrorists are attacking us by spreading disease with our money. Dollar bills are passed hand-to-hand more than mail, more than menus, more than a few door handles in an office building. No one gives a second thought if you handle money with gloves in the winter. Do you ever think twice when Achmed hands you your change at the convenience store? Now you know why the flu reached epidemic proportions this year.

Suzy: Yes, yes, that could be true. And it sure explains a lot. I'm going to be real careful taking any money from Muslims now.

Analysis Tom's conspiracy theory is just feeding Suzy's prejudices and paranoia. What's possible isn't necessarily plausible.

An interesting story is just that—a story, which might be worth investigating. We need evidence before we believe. Sometimes there really are conspiracies, like when the soldiers and Department of Defense tried to cover up the torture at Abu Ghraib. And with conspiracies, we can be pretty sure evidence will eventually come out.

> "Three may keep a secret, if two of them are dead."
>
> Benjamin Franklin

Similar mistakes in evaluating arguments

It's a mistake to say an argument is bad because of who said it. That's *mistaking the person (or group) for the argument.*

Example 22 Zoe: I went to Professor Zzzyzzx's talk about writing last night. He showed why the best way to start on a novel is to make an outline of the plot.

Suzy: Are you kidding? He could never get his published. And he doesn't even speak English good.

Analysis Suzy is mistaking the person for the argument. Professor Zzzyzzx's argument may be good even if Suzy doubts his qualifications to make it.

To *refute* an argument is to show it's bad. When someone points out to us that the person who made an argument doesn't believe one of the premises, we reckon the argument must be bad. But that's a *phony refutation*. Sincerity is not one of the criteria for an argument to be good. Judging by sincerity is mistaking the person for the argument.

Example 23 Harry: We should stop logging old-growth forests. There are very few of them left in the U.S. They are important watersheds and preserve wildlife. And once cut, we can't recreate them.

Tom: You say we should stop logging old-growth forests? Who are you kidding? You just built a log cabin on the mountain.

Analysis Tom's rejection of Harry's argument seems reasonable, since Harry's actions betray the conclusion he's arguing for. But whether they do or not (perhaps the logs came from the land that Harry's family cleared in a new-growth forest), Tom has not answered Harry's argument. Tom is not justified in ignoring an argument because of what he thinks Harry did.

If Harry responds to Tom by saying that the logs for his home weren't cut from an old-growth forest, he's been suckered. Tom got him to change the subject, and they will be debating an entirely different claim than Harry intended. It's a phony refutation.

Whether a claim is true or false
is not determined by who said it.

Whether an argument is good or bad
is not determined by who made it.

"First, realize that it is necessary for an intelligent person to reflect on the words that are spoken, not the person who says them. If the words are true, he will accept them whether he who says them is known as a truth teller or a liar. One can extract gold from a clump of dirt, a beautiful narcissus comes from an ordinary bulb, medication from the venom of a snake."
                    Abd-el-Kader, Algerian Muslim statesman, 1858

Always ask "Why?"
Always ask "So?"
Take as authority only those whose speech indicates knowledge and awareness and whose conduct indicates trustworthiness. Rely never on the position of an authority: many fools have been promoted to high place. Human desires, wills, fears can lead to fools prospering. But wisdom will out.

Don't believe because it's comfortable. A great desire for comfort, for no challenge, can lead to the enslavement of the truth and to the enslavement of us all.

If in doubt, suspend judgment. The seeker is wiser than the dogmatist.

6 Repairing Arguments

We need to repair arguments

Most arguments we encounter are not complete. But with a bit of thought we can see that lots of incomplete arguments can be good.

Example 1 Lee: Tom wants to get a dog.
  Maria: What kind?
  Lee: A dachshund. And that's really stupid, since he wants one that will catch a Frisbee.

Analysis Lee's made an argument: Tom wants a dog that will catch a Frisbee, so he shouldn't get a dachshund. This looks bad because there's no *glue*, no claim that gets us from the premise to the conclusion. But Maria knows, just like us, that a dachshund is a lousy choice for someone who wants a dog to catch a Frisbee. They're too low to the ground, they can't run fast, they can't jump, and the Frisbee is bigger than they are, so they couldn't bring it back. Any dog like that is a bad choice for a Frisbee partner. Lee just left out these obvious claims, but why should he bother to say them?

Folks usually leave out so much that if we look at only what's said in evaluating what we should believe, we'll be missing too much. We can and should repair many arguments. But when are we justified in adding a premise? How do we know whether we've repaired or just added our own ideas? And how can we recognize when an argument is beyond repair? We first need to make some assumptions about the person who's making the argument.

The Principle of Rational Discussion We assume that the other person who is discussing with us or whose argument we're reading:
- Knows about the subject under discussion.
- Is able and willing to reason well.
- Is not lying.

Why should we invoke this principle? After all, not everyone fits these conditions all the time.

Consider condition (1). Dr. E leaves his car at the repair shop because it's running badly, and he returns later in the afternoon. The mechanic tells him that he needs a new fuel injector. Dr. E asks, "Are you sure I need a new one?" That sounds like an invitation for the mechanic to give an argument. But she shouldn't. Dr. E doesn't have the slightest idea how his engine runs, so she might as well be speaking Greek. She should try to educate Dr. E or ask Dr. E to accept her claim on trust—after all, she's an authority.

Consider condition (2). Sometimes people intend not to reason well. Like the demagogic politician or talk-show host, they want to convince you by other means and will not accept your arguments, no matter how good they are. There's no point in deliberating with them.

Or you may encounter a person who is temporarily unable or unwilling to reason well, a person who's upset or in love. Again, it makes no sense at such a time to try to reason. Calm them, address their emotions, and leave discussion for another time.

Then again, you might find yourself with someone who wants to reason well but just can't seem to follow an argument. Why try to reason? Give them a copy of this book.

What about condition (3)? If you find that the other person is lying —not just a little white lie, but continuously lying—there's no point in reasoning with them, except perhaps to catch them in their lies.

The Principle of Rational Discussion does not instruct us to give other people the benefit of the doubt. It summarizes the necessary conditions for us to be reasoning with someone. Compare it to playing chess: what's the point of playing with someone if he doesn't understand or won't play by the rules?

Still, most people don't follow this principle. They don't care if your argument is good. Why should you follow these rules and assume them of others? If you don't:

- You're denying the essentials of democracy.
- You'll undermine your own ability to evaluate arguments.
- You're not as likely to convince others.

"If you once forfeit the confidence of your fellow citizens, you can never regain their respect and esteem. It is true that you may fool all the people some of the time; you can even fool some of the people all the time; but you cannot fool all of the people all the time."   Abraham Lincoln

Example 2 Dick (to Suzy): Cats are really dangerous pets. Look at all the evidence. See, it says so here in this medical journal, and it lists all the diseases you can catch from them, even schizophrenia they now think. You know that lots of your friends get sick from cat allergies. And remember how Puff scratched Zoe last week? You can't deny it.

Suzy: OK, OK, I believe what you've said. So you can reason well like Dr. E. But I still don't believe that cats are dangerous pets.

Analysis Suzy recognizes that Dick has given a good argument for cats being dangerous pets, but she still doesn't believe it. She's not judiciously suspending judgment; she's just unwilling to reason about her beloved cats.

The Mark of Irrationality If someone recognizes that an argument is good, then he or she is irrational not to accept its conclusion.

It's not worthwhile to reason with people if they're irrational. Sometimes, though, we hear an argument for one side and then one for the other, and we can't find a flaw in either. Then we should suspend judgment on which conclusion is true until we can investigate more. It's not irrational to suspend judgment if you're not sure.

The Guide to Repairing Arguments

The Principle of Rational Discussion can help us formulate a guide to repairing arguments.

The Guide to Repairing Arguments Given an (implicit) argument that is apparently defective, we are justified in *adding* a premise or conclusion if:

1. The argument becomes valid or stronger.
2. The premise is plausible and plausible to the other person.
3. The premise is more plausible than the conclusion.

We can also *delete* a premise that's false or dubious if doing so makes the argument no worse.

Example 3 Lee: I was wondering what kind of pet Louis has. It must be a dog.

Maria: How do you know?

Lee: Because I heard it barking last night.

Analysis Maria shouldn't dismiss Lee's reasoning just because the link from premises to conclusion is missing. She should ask what claim(s) are needed to make it strong, since by the Principle of Rational Discussion we assume Lee intends to and is able to reason well. The obvious premise to add is "All pets that bark are dogs." But Maria knows that's false (seals, foxes, parrots) and can assume that Lee does, too, since he's supposed to know about the subject. So she tries "Almost all pets that bark are dogs." That's plausible, and with it the argument is strong and good.

We first try to make the argument valid or strong, because we don't need to know what the speaker was thinking in order to do so. Then we can ask whether that claim is plausible and whether it would be plausible to the other person. *By first trying to make the argument valid or strong, we can show the other person what he or she needs to assume to make the argument good.*

Example 4 No dog meows. So Spot does not meow.

Analysis "Spot is a dog" is the only premise that will make this a valid or strong argument. So we add that. Then, if this new claim is plausible, the argument is good. We don't add "Spot barks." That's true and may seem obvious to the person who stated the argument, but it doesn't make the argument any better. So adding it violates condition (1) of the Guide. *We repair only as needed.*

Example 5 Almost every dog barks. So Spot is a dog.

Analysis The obvious premise to add is "Spot barks." That may be true, but it still leaves the argument weak: Spot could be a fox, or a seal, or a coyote. *If the obvious premise to add leaves the argument weak, the argument is unrepairable.*

Example 6 Dr. E is a good teacher because he gives fair exams.

Analysis A premise needed to make the argument strong is "Almost any teacher who gives fair exams is a good teacher." But that's dubious, since a bad teacher could copy fair exams from the instructor's manual. *The argument can't be repaired because the obvious premise to add to make the argument strong or valid is false or dubious.*

But can't we make it strong by adding, say,"Dr. E gives great explanations," "Dr. E is amusing," "Dr. E never misses class," . . . ? Yes, all those are true and perhaps obvious to the person. But adding them doesn't repair this argument. It makes a whole new argument. *Don't put words in someone's mouth.*

Example 7 Sure you'll get a passing grade in English. After all, you paid tuition to take the course.

Analysis The argument is weak—and it *is* an argument: the last sentence is meant as reason to believe the first. But there's no obvious repair: it's false that anyone who pays tuition for a course will pass it. *The person apparently can't reason.* Don't bother to repair this one.

Example 8 You shouldn't eat bacon. Haven't you heard that fat is bad for you?

Analysis The conclusion is the first sentence. But what are the premises? The speaker's question is rhetorical, meant to be taken as an assertion: "Fat is bad for you." That alone, though, won't give us the conclusion. We need something like "Bacon has a lot of fat" and "You shouldn't eat anything that's bad for you." Premises like these are so obvious we don't bother to say them. This argument is OK with these obvious additions.

Example 9 Dick: Dogs are loyal. Dogs are friendly. Dogs can protect you from burglars.
    Maria: So?
    Dick: So dogs make great pets.
    Maria: Why does that follow?

Analysis Maria's right. Dick's argument is missing the *glue, the link between premises and conclusion that rules out other possibilities*, in this case something like "Anything that's loyal, friendly, and can protect you from burglars is a great pet." But that's exactly what Maria thinks is false: Dogs need room to run around, they need to be walked every day, they cost more to take care of than goldfish. *Just stating a lot of obvious truths doesn't by itself get you the conclusion.*

Example 10 You're going to vote for the Green Party candidate for President? Don't you realize that means your vote will be wasted?

Analysis Here, too, the questions are rhetorical, meant to be taken as assertions: "You shouldn't vote for the Green Party candidate" (the conclusion) and "Your vote will be wasted" (the premise). This sounds reasonable, though something is missing. A visitor from Denmark may not know that "The Green Party candidate doesn't have a chance of winning" is true. But she may also question why that matters. We'd have to fill in the argument further: "If you vote for someone who doesn't have a chance of winning, then your vote will be wasted." And when we add that premise, we see that the argument which uses such "obvious" premises is really not good. Why should we believe that if you vote for someone who doesn't stand a chance of winning then your vote's wasted? If that were true, then who wins is the only important result of an election, rather than, say, making a position understood by the electorate for the next election. At best, we can say that when the unstated premises are added, we get an argument one of whose premises is in need of a substantial argument. *Trying to repair an argument can lead us to unstated assumptions that need to be debated.*

Example 11   Cats are more likely than dogs to carry diseases harmful to humans. Cats kill songbirds and can kill people's pets. Cats disturb people at night with their screeching and clattering in garbage cans. Cats leave paw prints on cars and will sleep in unattended cars. Cats are not as pleasant as dogs and are owned only by people who have satanic affinities. So there should be a leash law for cats just as much as for dogs.

  Analysis   This letter to the editor is going pretty well until the next to last sentence. *That claim is a bit dubious and the argument would be just as strong without it. So we should delete it.* Then we have an argument which, with some unstated premises you can supply, is pretty good.

Example 12   Alcoholism is a disease, not a character flaw. People are genetically predisposed to be addicted to alcohol. An alcoholic should not be fired or imprisoned, but should be given treatment. Treatment centers should be established, because it is too difficult to overcome the addiction to alcohol by oneself. The encouragement and direction of others is what's needed to help people, for alcoholics can find the power within themselves to fight and triumph over their addiction.

  Analysis   On the face of it, "Alcoholism is a disease, not a character flaw" and "Alcoholics can find the power within themselves to fight and triumph over their addiction" contradict each other. Since both are used to get the conclusion, "Treatment centers should be established," neither can be deleted. They both can't be true, so the argument is unrepairable.

Example 13 "In a famous speech, Martin Luther King Jr. said:

> I have a dream that one day this nation will rise up and live out
> the true meaning of its creed: 'We hold these truths to be self-
> evident—that all men are created equal'. . . . I have a dream
> that one day even the state of Mississippi, a desert state swelter-
> ing with the heat of injustice and oppression, will be transformed
> into an oasis of freedom and justice. I have a dream that my four
> little children will one day live in a nation where they will not be
> judged by the color of their skin but by the content of their
> character.

. . . King is also presenting a logical argument . . . the argument might
be stated as follows: 'America was founded on the principle that all
men are created equal. This implies that people should not be judged
by skin color, which is an accident of birth, but rather by what they
make of themselves ('the content of their character'). To be consistent
with this principle, America should treat black people and white people
alike.' "
David Kelley, The Art of Reasoning

Analysis The rewriting of this passage is putting words in
someone's mouth. Where did David Kelley get the premise "This
implies . . . " ? Stating my dreams and hoping others will share them
is not an argument. Martin Luther King Jr. knew how to argue well
and could do so when he wanted. We're not going to make his words
more respectable by pretending they're an argument. *Not every good
attempt to persuade is an argument.*

Example 14

Analysis Tom has confused whether we have the right to cut
down forests with whether we should cut them down. The argument is

weak; indeed, we could delete either premise and it wouldn't be any weaker. That's just to say that his premises are irrelevant. A premise is *irrelevant* if you can delete it and the argument isn't any worse.

Example 15   "U.S. citizens are independent souls, and they tend to dislike being forced to do anything. The compulsory nature of Social Security therefore has been controversial since the program's beginnings. Many conservatives argue that Social Security should be made voluntary, rather than compulsory."

J. M. Brux and J. L. Cowen, Economic Issues and Policy

Analysis   The first two sentences look like an argument. But the first is too vague to be a claim, and there's no obvious way to make it precise. So we can't view this as an argument, and we shouldn't try to make it into one.

Example 16   Maria (to Dick): Parking is still difficult on campus. From 8:30 in the morning till 4 every afternoon it's impossible to find a parking place—I've spoken with all my friends, and it takes us 15 minutes, often more. And that's with a parking sticker we paid $25 to get. Without that you could look forever, or park on the street and hope not to get a ticket. Our school should build more parking lots.

Analysis   Dick agrees with all of Maria's assumptions. But he still asks, "So?" There are lots of ways those claims could be true and Maria's conclusion false: The school doesn't want to encourage more people to drive to campus; the school has no money to build parking lots; the school has an agreement with the city not to build more parking. *Some general claim is needed, but there's no obvious plausible one, so we can't repair the argument.*

Example 17   (a) Investors in 1997 invested more than twice as much money in no-load mutual funds as in other mutual funds. So, (b) investors in 1997 overwhelmingly preferred no-load mutual funds.

Analysis   Typically, we invoke some evidence such as (a), which is objective, to conclude (b), which is subjective. But to have a good argument for (b) we also need a premise like "When people invest money in a fund, they prefer that fund to one that they do not invest in," which is plausible and makes this a good argument. That subjective claim is the link between the observed behavior and the inferred state of mind. *Often an unstated assumption linking behavior to thoughts is needed to make an argument good.*

Example 18 None of Dr. E's students are going to beg in the street. 'Cause only poor people beg. And Dr. E's students will be rich because they understand how to reason well.

    Analysis This is a superb argument!

Unrepairable arguments

We've seen examples where it's clear that an argument is bad and there's no point trying to repair it. Let's summarize those conditions.

Unrepairable Arguments We can't repair an argument if any of the following hold:

- There's no argument there.
- There's nothing obvious to add.
- A premise is false or dubious and can't be deleted.
- The obvious premise to add would leave the argument weak.
- The obvious premise to add to make the argument valid or strong is not plausible.
- The conclusion is clearly false.

    But remember: *When you show that an argument is bad, you haven't shown that the conclusion is false.* A bad argument tells us nothing about the conclusion.

Implying and inferring

Example 19 Harry: I'm not going to vote, because no matter who becomes mayor, nothing is going to get done to repair roads in this part of town.

    Analysis An unstated claim is needed to make sense of what Harry said: "If no matter who becomes mayor nothing is going to get done to repair roads in this part of town, then you shouldn't vote for mayor." We infer this from what he said; Harry has implied it.

    When someone leaves a conclusion unsaid, he or she is ***implying*** it. When you decide that an unstated claim is the conclusion, you're ***inferring*** that claim. We also say someone is implying a claim if in context it's clear that he or she believes it. In that case we infer that the person believes the claim.

Example 20   Maria: Why did you write up this exercise?
                       It wasn't assigned.
   Lee: Dr. E said that all his best students hand in the optional
            exercises for extra credit.
   Maria: I better do one, too.

Analysis   Dr. E hasn't said that his students should hand in extra
work to get a good grade. But Lee and Maria have inferred that; they
think he's implied it.

Example 21   If Lee complains to the department head that Dr. E is
demanding more than he asked on the syllabus, Dr. E could reply that
Lee was jumping to conclusions. He might say, "I've observed that my
best students hand in extra-credit work—that's all I was saying. I had
no intention of asking for extra assignments." Lee, however, could say
that in the context in which Dr. E made the remark it was fairly obvious
he was implying that if Lee wanted him to believe he's a good student,
he should hand in extra work. Implying and inferring can be risky.

Example 22   "A member of Pakistan's parliament stood his ground in
August, defending news reports from his Baluchistan province that five
women had been shot and then buried alive as tribal punishment for
objecting to their families' choosing husbands for them. A defiant Israr
Ullah Zehri told the Associated Press, 'These are centuries-old traditions, and I will continue to defend them,' despite condemnation by
Zehri's colleagues. 'Only those who indulge in immoral acts should be
afraid,' Zehri said."   New York Daily News-AP, 8/30/08

   Analysis   We can infer from what Zehri said that he believes if an
act is in accord with a centuries-old tradition, it is morally acceptable.
He has implied that.

7 Counterarguments

Counterarguments

Raising objections and answering other people's objections is an important part of making and evaluating arguments.

Example 1 Dick: Zoe, we ought to get another dog.
    Zoe: What's wrong with Spot?
    Dick: Oh, no, I mean to keep Spot company.
    Zoe: Spot has us. He doesn't need company.
    Dick: But we're gone a lot. And he's always escaping from the yard, 'cause he's lonely. And we don't give him enough time. He should be out running around more.
    Zoe: But think of all the work! We'll have to feed the new dog. And think of all the time necessary to train it.
    Dick: I'll train him. We can feed him at the same time as Spot, and dog food is cheap. It won't cost much.

Analysis Dick is trying to convince Zoe to believe "We should get another dog." But he has to answer her objections:

We ought to get another dog.
    (*objection*) We already have Spot.
The other dog will keep Spot company. (*answer*)
    (*objection*) Spot already has us for company.
We are gone a lot. (*answer*)
He is always escaping from the yard. (*answer*)
He's lonely. (*answer*)
We don't give him enough time. (*answer*)
He should be out running around more. (*answer*)
    (*objection*) It will be a lot of work to have a new dog.
    (*objection*) We will have to feed the new dog.
    (*objection*) It will take a lot of time to train the new dog.
I (Dick) will train him. (*answer*)
We can feed him at the same time as Spot. (*answer*)
Dog food is cheap. (*answer*)

Argument. Counterargument. Counter-counterargument. Objections are raised: Someone puts forward a claim that, if true, shows that one of our claims is false or at least doubtful or that our argument is not valid or strong. We then have to answer that challenge to sustain our argument. Knocking off an objection is a mini-argument within your argument; if it's not a good (though brief) one, it won't do the job.

But reasoning well isn't about winning. You could say, "I hadn't thought of that, I guess you're right." Or, "I don't know, I'll have to think about that."

In making an argument, you'll want to make it strong. You might think you have a great argument. All the premises seem obvious and they lead to the conclusion. But if you imagine someone objecting, you can see how to give better support for doubtful premises or make it clearer that the argument is valid or strong. When you answer counter-arguments in your own writing, it allows the reader to see you haven't ignored some obvious objections. Just make a list of the pros and cons. Then answer the other side.

Refuting an argument

It's useless to kill flies. The ones you kill will be the slowest, because the fast flies will evade you.

So you will be killing off the slowest ones and the fastest ones will remain. Over time, then, the genes for being fast will predominate.

Then with super-fast flies, it will be impossible to kill them anyway. So it's useless to kill flies.

Zoe can't let it pass. But how do you refute an argument?

Zoe might object to one of the premises, saying Dick won't be killing the slowest but only the ones that happen to come into their house.

Or she could agree with the premises but note that "over time" could be thousands of years, so the conclusion doesn't follow.

Or she could attack the conclusion, saying that it's not useless to kill flies because she does it all the time and it keeps their home clean.

All the ways we can show an argument is unrepairable are useful in refuting an argument. Three are fundamental.

> ***Direct ways of refuting an argument***
> * Show that at least one of the premises is false or implausible.
> * Show that the argument isn't valid or strong.
> * Show that the conclusion is false.

Sometimes we can't point to any one premise that's false or dubious, but we know there's something wrong with the premises. They might get the conclusion that's argued for, but they get a lot more, too—so much that we can see the premises are inconsistent or lead to an absurdity.

> ***Reducing to the absurd*** To reduce to the absurd is to show that at least one of several claims is false or that collectively they are unacceptable by drawing a false or unwanted conclusion.

Example 2 Tom: Everyone in the U.S. should have to speak English. Everyone's got to talk the same, so we can communicate easily. And it'll unify the country.

Lee: Sure. But I have real trouble understanding people from New York. So we should make everyone speak just like me, from Iowa.

Analysis Lee is reducing Tom's argument to the absurd. Starting with the same premises he gets a claim he knows Tom won't accept.

Example 3 You complain that taxes are already too high and there's too much crime. And you say we should permanently lock up everyone who has been convicted of three felonies. In the places where this has been done it hasn't reduced the crime rate. So we will have many more people who will be in jail for their entire lives. We'll need more prisons, many more. We'll need to employ more guards. We'll need to pay for a lot of health care for these people when they get old. So if you lock up everyone who has been convicted of three felonies, we'll have to pay substantially higher taxes. Since you're adamant that taxes are too high, you should abandon your claim that we should lock up forever everyone who's been convicted of three felonies.

Analysis The speaker is showing that the claim that taxes are too high and shouldn't be higher contradicts the claim that we should permanently lock up everyone who has been convicted of three felonies.

When you use this indirect method of refuting, be sure that the argument you use to get the false or absurd conclusion is good. Otherwise, it could be the claims you introduce that give the contradiction.

Example 4 Zoe: I can't believe you're eating those *Baken-ets* fried pork skins. I thought you wanted to lose weight.

Dick: Right, and the package says on the front in big letters "0 g Net Carbs."

Analysis Dick believes he's refuted the unstated claim that fried pork skins are fattening. But his refutation rests on an assumption that's false: If a food has no carbohydrates, it's not fattening.

One way to reduce to the absurd is to use similar premises in an argument that sounds just like the original yet leads to an absurd conclusion.

Example 5

LOOK, YOUR ARGUMENT AGAINST KILLING FLIES IS BAD. I COULD USE THE SAME ARGUMENT AGAINST KILLING BACTERIA, OR AGAINST KILLING CHICKENS FOR DINNER FROM AUNT MARGERY'S HENHOUSE. THOSE CONCLUSIONS WOULD BE ABSURD.

Example 6 "Guns in restaurants? Great idea, but why stop there? How about day care centers? Churches? Hospitals? High schools? Better yet, high schools during rival basketball games!

Now there's an idea!

Let's make guns legal in courtrooms, too. After all, it's a Second Amendment right. City Council and PTA meetings, too.

Finally, how about the halls of Congress? After all, the Second Amendment says '. . . the right of the people to keep and bear arms shall not be infringed.' So grab the Glock and go visit the politician of choice in his or her office."

Bill Dobbeck, letter to the editor, *Albuquerque Journal*, 2/4/10

Analysis In the debate in New Mexico over whether people should be allowed to carry guns into restaurants, little was given as justification for that other than Second Amendment rights and people's rights to do as they wish. The author refutes those by reducing to the absurd.

Example 7   You say we should leave you alone and let you cockfight because it's a tradition of your New Mexican Hispanic culture? Well, arranged marriages for 12-year-old girls were a tradition in some parts of this country. So was wife beating. We stopped those because, like cockfighting, they're cruel.

Analysis   This refutation by analogy goes further by giving a general claim that would sanction the opposite of the conclusion: We should stop traditions that are cruel.

There are several bad ways to try to refute. We've already seen *phony refutations*. The worst is **ridicule**, which ends a discussion, belittles others, and makes enemies.

Example 8   Dr. E: I hear your department elected a woman as chairman.
Prof. Smythe: Yes, indeed. And now we're trying to decide what we should call her—"chairman," or "chairwoman," or "chairperson."
Dr. E: "Chairperson"? Why not use a neutral term that's really appropriate for the position, like "chaircreature"?

Analysis   No argument has been given for why "chairman" shouldn't be replaced by "chairperson," although Dr. E thinks he's shown the idea is ridiculous.

In rational discussion ridicule is a worthless device: It ends arguments, belittles the other person, and makes enemies. In theory there's a big difference between reducing to the absurd and ridicule. But in practice it's difficult to distinguish them. Often not enough of an argument is given to see how the absurd conclusion follows, so it sounds like ridicule. *If someone wants us to see his comments as an argument, it's his responsibility to make that clear.* Otherwise, let's classify it as ridicule.

REDUCE TO THE ABSURD

{A, B, ..., C} — THE OTHER PERSON'S CLAIMS

(OTHER PLAUSIBLE CLAIMS)

D - CONCLUSION, FALSE OR ABSURD

RIDICULE

A — THE OTHER PERSON'S CLAIM

HA! HA! HA! HA! HA!

Finally, there's the all-purpose way to evade another person's argument by giving a *strawman*: putting words in someone's mouth.

Example 9   Tom: Unless we allow the logging of old-growth forests in this county, we'll lose the timber industry and these towns will die.

Dick: So you're saying that you don't care what happens to the spotted owl and to our rivers and the water we drink?

Tom: I said nothing of the sort. You've misrepresented my position.

Analysis   The only reasonable response to a strawman is to say calmly, "That isn't what I said."

8 Concealed Claims

Sometimes people try to get us to assume a claim is true without reflecting on it. A *slanter* is an attempt to convince by using words that conceal a claim which is dubious.

Persuasive definitions are slanters. So, too, are *loaded questions* which presuppose that some dubious claim is true.

Example 1 Lee: Why can't cats be taught to heel?
Suzy: What makes you think that cats can't be taught to heel?
Analysis Lee poses a loaded question. Suzy answers it by pointing out and challenging the concealed claim.

Example 2 Dick: Why do all women like to shop?
Zoe: We don't.
Analysis Zoe has answered Dick's loaded question by denying his concealed claim.

A *euphemism* is a word or phrase that makes something sound better than a neutral description; a *dysphemism* makes it sound worse.

Example 3 Suzy: You should try to fix up Wanda with a date. Tell your friends she's Rubenesque.
Tom: You mean she's fat.
Analysis Suzy's used a euphemism.

Example 4 The freedom fighter attacked the convoy.
Analysis "Freedom fighters" is a euphemism, concealing the claim that the guerillas are good people fighting to liberate their country and give their countrymen freedom.

Example 5 The terrorists attacked the convoy.
Analysis "Terrorist" is a dysphemism, concealing the claim that the guerillas are bad people, inflicting violence on civilians for their own partisan ends without popular support.

Example 6 The merciless slaughter of seals for their fur continues in a number of countries.
Analysis "Merciless slaughter" is a dysphemism; "harvesting" would be a euphemism, while "killing" would be a neutral description.

Example 7 "American authorities suffered their own black eye over mistreatment of prisoners when photographs surfaced early last year showing U.S. soldiers abusing detainees at the Abu Ghraib prison on Baghdad's western outskirts." Associated Press, 12/26/2005

Analysis The bias in this newspaper account is clear: "black eye" is a euphemism for serious damage to their reputation and "abusing" is a euphemism for torture.

Example 8 The homepage of the website of Los Alamos National Laboratory has the following items you can click on: Science and Technology, Working with LANL, Organization, Community, Education & Internships, Life@LANL, International.

Analysis The homepage is a euphemism by *misdirection.* If you didn't already know that LANL is one of the main centers for nuclear weapons research in the United States, you might think from this that it is devoted solely to scientific research and its applications for the betterment of humanity.

A *downplayer* is a word or phrase that minimizes the significance of a claim; an *up-player* exaggerates.

Example 9 Zoe: Hey Mom. Great news. I managed to pass my first French exam.

Mom: You only just passed?

Analysis Zoe has up-played the significance of what she did, concealing the claim "It took great effort to pass" with the word "managed." Her mother downplayed the significance of passing by using "only just," concealing the claim "Passing and not getting a good grade is not commendable."

Example 10 "New Mexico surveys show 60 percent of high schoolers have had sex before graduating, and only 12 percent remain abstinent until marriage." *Albuquerque Journal*, 1/13/2005

Analysis "Only"? Why shouldn't we be surprised it's so many?

A *weaseler* is a claim that's qualified so much that the apparent meaning is no longer there.

Example 11 If you buy *The Pocket Guide to Critical Thinking*, you'll get a job paying 25% more than the average wage in the U.S.*

---

* Purchaser agrees to study this book four hours per day for two years.

Example 12 "[Elliot] Rappaport [a forensic psychologist] said a psychological autopsy is just like any other psychological evaluation except that the patient is missing." *Albuquerque Journal*, 12/3/2004

Analysis That "except" qualifies the comparison away: dogs are just like cats, except that dogs are canines and cats are felines.

Example 13 Maria (to her boss): I am truly sorry it has taken so long for you to understand what I have been saying.

Analysis Maria has not apologized.

A *proof substitute* is a way to convince by suggesting you have a proof without actually offering one.

Example 14 Dr. E to Suzy: Cats can't reason. It's obvious to any thinking person. Being around them so much must have convinced you of that. Of course some people are misguided by their emotions into thinking that felines have intelligence.

Analysis Dr. E didn't prove that cats can't reason, though he made it sound as if he were proving something. He was just reiterating the claim, trying to browbeat Suzy into believing it with the words "obvious," "must have convinced," "some people are misguided."

Example 15 Suzy: Cats can so reason. It's been shown over and over that they can.

Analysis Unless Suzy can point to some studies, this is a proof substitute, too.

Ridicule is a particularly nasty form of proof substitute: that's so obviously wrong it's laughable.

Example 16 Dr. E: Cats can reason? Sure, and the next thing you know you'll be inviting them over to play poker.

Another way to conceal that you have no support for your claim is to *shift the burden of proof*.

Example 17 Tom: The university should lower tuition.
Maria: Why?
Tom: Why not?

Analysis Tom hasn't given any reason to think his claim is true. He's only invited Maria to say why she thinks it's false, so he can attack that—which is easier than supporting his position.

Example 18   Why wouldn't you want laws limiting where sexual predators can live?
  Analysis   Often an attempt to shift the burden of proof is given as a question that assumes a default judgment.

  Here are some more examples of how people conceal claims.

Example 19   Scientists at McEpstein University's School of Medicine have discovered a cure for baldness.
  Analysis   This supposes that baldness is a disease, or at least a disability that needs to be cured.

Example 20   Tom: This book *Esperanza's Box of Saints* by Maria Escandón is great. It's so good that if I didn't know better I would have thought it was written by a man.
  Analysis   Tom hasn't concealed very well his low opinion of women writers.

Example 21   Blondes aren't dumb—they're just slow
"BERLIN—Blonde women are not dumber than brunettes or redheads, a reassuring study shows—they are just slower at processing information, take longer to react to stimuli and tend to retain less information for a shorter period of time than other women.
  'This should put an end to the insulting view that blondes are airheads,' said Dr. Andrea Stenner, a blonde sociologist who studied more than 3,000 women for her doctoral research project."

                                                  Weekly World News, 10/15/1996
  Analysis   This is an example of saying it isn't so while showing it is, hoping no one will notice.

Example 22   Wages for the same kind of labor are lower in the South than in the North. Also, wages are lower in Puerto Rico than in the United States. How can a northern employee protect his wage level from the competition of lower-wage southern labor? And how can a laborer in the United States protect his job (and higher wage rate) from Puerto Rican labor? One device would be to advocate "equal pay for equal work" in the United States, including Puerto Rico, by legislating minimum wages higher than the prevailing level in the South and Puerto Rico. It should come as no surprise to learn that in the United States support for minimum-wage laws comes primarily from northerners who profess to be trying to help the poorer southern
laborers.   A. Alchian and W. Allen, *University Economics*

Analysis The word "profess" here conceals the claim that northern workers—on the whole—are duplicitous: they say they're trying to help their southern counterparts when they're really motivated by self-interest. But the authors have given no reason to believe that those workers are not sincere. By noticing how the authors have used a slanter here, we can be on the alert for this bias against workers in their book.

You should avoid concealing claims because the other person can destroy your points not by facing your reasoning, but by pointing out how you are trying to confuse people. If you reason calmly and well, you will earn the respect of others and may learn that others merit your respect, too.

9 Fallacies

Some arguments are typically bad because they use a premise that's false or would usually be dubious. For example, a drawing the line fallacy requires as premise "If you can't make the difference precise, then there's no difference." A subjectivist fallacy requires "If there is a lot of disagreement about whether this claim is true, then the claim is subjective."

Other mistakes in reasoning come from violating the Principle of Rational Discussion—for example, shifting the burden of proof or begging the question.

Still other arguments are typically bad because of their structure involving words such as "if . . . then . . ." or "all," as we'll see in the next two chapters.

We label such typically bad arguments as *fallacies* in order to make the evaluation of them easier: all we need to do is note that the argument is a fallacy or, if it's of a kind that uses a typically dubious premise, show that in this case the premise actually is dubious, as with a bad appeal to authority. Labeling certain kinds of arguments as fallacies also serves as a warning against using them.

We'll see many more kinds of fallacies in the chapters that follow. Here we'll consider only some kinds that require a dubious premise that's meant to engage our emotions. Emotions do and should play a role in our reasoning: we cannot make good decisions if we don't consider their significance in our emotional life. But that doesn't mean we should be swayed entirely by our emotions. An *appeal to emotion* is an argument that depends on a premise that says you should believe or do something just because you feel a certain way.

Example 1 Maria: You should vote for Ralph for school president.
    Zoe: Why?
    Maria: Because he doesn't have many friends.
    Analysis To construe this as a strong or valid argument, we'd need to add "You should vote for someone you feel sorry for." That's an *appeal to pity*, and in this case it's implausible.

Example 2 Dick (to Zoe): We should give to the American Friends Service Committee. They help people all over the world help themselves, and they don't ask whether those folks agree with their principles. They've been doing it really well for nearly a century now, and they have very low overhead: almost all the money they get goes to those in need. All those people who don't have running water or health care need our help. Think of those poor kids growing up malnourished and sick. We've got enough money to send them $50.

Analysis This requires an unstated premise appealing to pity. But it isn't just "Do it just because you feel sorry for someone." Rather "If you feel sorry for people, *and* you have a way to help them that is efficient and morally upright, *and* you have enough money to help, then you should send the organization money" will make the argument strong. That seems plausible, though whether it is the best use of Zoe and Dick's money needs to be addressed.

Example 3 You shouldn't drive so quickly in the rain. The roads are very slippery after the first rain of the season and we could get into an accident.

Analysis Normally an appeal to emotion by itself is not sufficient to make a good argument. But sometimes an ***appeal to fear***, as in this example, can be a legitimate factor in making a decision. An appeal to emotion that concludes you should *do* something can be good or it can be bad.

Example 4 Wanda: This diet will work because I have to lose 20 pounds by the end of the month for my cousin's wedding.

Analysis This is an example of ***wishful thinking***, and it's bad. An appeal to emotion whose conclusion is a description of the world is going to be bad (unless you can delete that premise). Why should we believe that some description is true just because we're moved by our emotions? Wishing it so don't make it so.

It's not bad to appeal to emotions, but usually it is bad to appeal only to emotions in reasoning well.

Special Kinds
of Claims

10 Compound Claims

Compounds, contradictories, and "or" claims

> **Compound claim** A *compound claim* is one that is composed of other claims but has to be viewed as just one claim.

Example 1 Either a Democrat will win the election or a Republican will win the election.

   Analysis This is a compound claim composed of the claims "A Democrat will win the election" and "A Republican will win the election" joined by the word "or." Whether it's true or false depends on whether one or both of its parts are true, but the entire sentence is just one claim. The claims that make up an "or" claim are called the *alternatives*.

Example 2 Dick or Zoe will go to the grocery to get eggs.

   Analysis We can view this as an "or" claim with alternatives "Dick will go to the grocery to get eggs" and "Zoe will go to the grocery to get eggs."

Example 3 Lee will pass his exam because he studied so hard.

   Analysis This is not a compound claim: "because" is an indicator word that tells us this is an argument.

> **Contradictory of a claim** A *contradictory* of a claim is one that must have the opposite truth-value.

Example 4 Spot is barking.

   Analysis A contradictory of this is "Spot is not barking."

Example 5 Inflation will be not less than 3% this year.

   Analysis A contradictory of this is "Inflation will be less than 3% this year," which doesn't contain "not."

   In order to discuss contradictories and some arguments that depend on the forms of claims, it's useful to have the following conventions in discussions and diagrams:

The letters *A*, *B*, and *C* stand for any claims.

The phrase "not *A*" stands for "a contradictory of *A*".

An arrow ↓ stands for "therefore".

The symbol + means an additional premise.

Contradictory of an "or" claim

A or B has contradictory *not A and not B*

Contradictory of an "and" claim

A and B has contradictory *not A or not B*

Example 6 Maria got the van or Manuel won't go to school.

Analysis A contradictory is "Maria didn't get the van, and Manuel will go to school."

Example 7 Tom or Suzy will pick up Manuel for class today.

Analysis A contradictory is "Neither Tom nor Suzy will pick up Manuel for class today." We can use "neither . . . nor . . ." for a contradictory of an "or" claim.

We can recognize some arguments as valid from just their form.

Excluding possibilities

   A or B + *not A*
   ――――――――――   *Valid*
        ↓
        B

Example 8 Either there's a wheelchair ramp at the school dance, or Manuel stayed home. But there isn't a wheelchair ramp at the school dance. So Manuel stayed home.

Analysis This is valid.

Example 9 Either all criminals should be locked up forever, or we should put more money into rehabilitating criminals, or we should accept that our streets will never be safe, or we should have some system for monitoring ex-convicts. [*This is all one claim*: *A* or *B* or *C* or *D*.] We can't lock up all criminals forever, because it would be too expensive. We definitely won't accept that our streets will never be safe. So either we should put more money into rehabilitating criminals, or we should have some system for monitoring ex-convicts.

Analysis Arguments that exclude just one or more of many alternatives are valid, too.

But remember: not every valid argument is good.

> **False dilemma**   A *false dilemma* is a bad use of excluding possibilities where the "or" claim is false or dubious. Sometimes just the dubious "or" claim is called the false dilemma.

Example 10

Analysis   Zoe poses a false dilemma: they could economize by not buying fast food lunches or by driving less.

Example 11   "Society can choose high environmental quality but only at the cost of lower tourism or more tourism and commercialization at the expense of the ecosystem, but society must choose. It involves a tradeoff."   R. Sexton, *Exploring Economics*

Analysis   This is a false dilemma. Costa Rica has created a lot of tourism by preserving almost 50% of its land in parks. When you see a *versus*-claim, suspect a false dilemma.

Example 12   "The question is whether the Taliban are ready to help build a 21st Century Afghanistan or whether they just want to kill people."   Secretary of Defense Robert Gates, Bloomberg News, 1/23/10

Analysis   This would be a false dilemma if "a 21st Century Afghanistan" weren't too vague for this to be a claim.

Conditional claims

> **Conditional claim**   A *conditional claim* is a claim that is or can be rewritten as one of the form *If A then B* that must have the same truth-value. The claim *A* is the **antecedent** and the claim *B* is the **consequent**.

Example 13   If Spot ran away, then the gate was left open.

Analysis This is a conditional, with antecedent "Spot ran away" and consequent "The gate was left open." The consequent need not happen later.

Example 14 I'll never talk to you again if you don't apologize.
    Analysis This is a conditional with antecedent "You don't apologize" and consequent "I'll never talk to you again."

Example 15 Loving someone means you never throw dishes at him.
    Analysis This is a conditional with antecedent "You love some-one" and consequent "You never throw dishes at him." It's not a defi-nition even though it uses the word "means."

Example 16 A mammal is an ungulate if it has hoofs.
    Analysis This is not a conditional or a compound. It's a defini-tion that uses "if" instead of "means that." We have to use our judg-ment to decide whether a claim is a conditional.

Example 17 If Dick goes to the basketball game, then either he got a free ticket or he borrowed money for one.
    Analysis This is a conditional whose consequent is also a compound claim.

Contradictory of a conditional
If A, then B has contradictory *A but not B*

    Zoe: I'm so worried. Spot got out of the yard. If he got out of the yard, then the dogcatcher got him, I'm sure.
    Suzy: Don't worry. I saw Spot. He got out of the yard, but the dogcatcher didn't get him.

If    *then*

contradictory:

but

A contradictory of a conditional is *not* another conditional.

Example 18   If Spot barks, then Suzy's cat will run away.
      Contradictory   Spot barked, but Suzy's cat did not run away.

Example 19   If Spot got out of the yard, he was chasing a squirrel.
      Contradictory   Spot got out of the yard, but he wasn't chasing a squirrel.

Example 20   If cats had no fur, they wouldn't give people allergies.
      Contradictory   Even if cats had no fur, they would still give people allergies. "Even if" is often used to make a contradictory where there's a false antecedent. The "if" in it does not create a conditional.

Example 21   Bring me an ice cream cone and I'll be happy.
      Contradictory: Despite that you brought me an ice cream cone, I'm not happy. "Despite that" is also used to make a contradictory of a conditional.

Example 22   (†) If Suzy handed in all her English homework, then she passed the course.
      Analysis   A contradictory is: Suzy handed in all her English homework, but she didn't pass the course."

The following are *not* contradictories:

"If Suzy didn't hand in all her English homework, then she passed the course."

(Both this and † could be true if Suzy passed anyway.)

"If Suzy handed in all her English homework, then she didn't pass the course."

(Both this and † could be true if Suzy didn't hand in all her homework.)

"If Suzy didn't hand in all her English homework, then she didn't pass the course."

(Both this and † could be true.)

Necessary and sufficient conditions

Two types of claims are closely related to conditionals.

Contrapositive of a conditional

The *contrapositive* of *If A, then B* is *If not B, then not A.*

The contrapositive is true exactly when the original is true.

Example 23 If Zoe does the dishes, then Dick will walk Spot.

Contrapositive If Dick doesn't walk Spot, then Zoe didn't do the dishes.

"Only if" claims

A only if B means the same as *If not B, then not A.*

"Only if " does not mean the same as "if."

Example 24 Dick will go into the army only if there is a draft.

Analysis This example means the same as "If there is no draft, then Dick will not go into the army."

Example 25 The following are equivalent:

You'll get a speeding ticket only if you're going over the speed limit.

If you're not going over the speed limit, then you won't get a speeding ticket.

If you get a speeding ticket, then you went over the speed limit.

Necessary and sufficient conditions

A is necessary for B means that *If not A, then not B* must be true.

A is sufficient for B means that *If A, then B* must be true.

Example 26 Passing an eye test is necessary but not sufficient for getting a driver's license.

   Analysis This means "If you don't pass an eye test, you can't get a driver's license" is always true; but "If you pass an eye test, you get a driver's license" need not be true.

Example 27 You can pass calculus only if you study hard.

   Analysis This isn't the same as "If you study hard, you can pass calculus." Rather, studying hard is necessary, required to pass calculus; it's not sufficient. The example is equivalent to "If you pass calculus, then you studied hard." Confusing "only if " with "if " is confusing a necessary with a sufficient condition.

Some valid forms of arguments with conditionals

There are some simple valid forms of reasoning using conditionals that are useful for analyzing and making arguments, and some similar arguments that are usually weak. Here's an illustration for some.

If Flo comes over to play,

If it's the day for the garbageman,

Then Dick will wake up.

If Suzy calls early,

If Spot barks,

The direct way of reasoning with conditionals	Affirming the consequent
$\dfrac{\textit{If A, then B } + A}{B}$ *Valid*	$\dfrac{\textit{If A, then B } + B}{A}$ *Weak*

Example 28 If Spot barks, then Dick will wake up. Spot barked. So Dick woke up.

Analysis This is a valid argument. It is impossible for both the premises to be true and conclusion false. It's an example of the direct way of reasoning with conditionals.

Example 29 If Spot barks, then Dick will wake up. Dick woke up. So Spot barked.

Analysis This is weak. Maybe Suzy called, or Flo came over to play. It's affirming the consequent, reasoning backwards.

Example 30 Wanda: If I go on Jane Fonda's workout and diet plan, I'll lose weight.
later Zoe: Did you see how much weight Wanda lost?
Suzy: She must have gone on that workout plan by Jane Fonda.

Analysis Suzy's overlooking other possibilities. Maybe Wanda has become bulimic or had liposuction. Affirming the consequent is reasoning backwards.

The indirect way of reasoning with conditionals	Denying the antecedent
$\dfrac{\textit{If A, then B } + \textit{not B}}{\textit{not A}}$ *Valid*	$\dfrac{\textit{If A, then B } + \textit{not A}}{\textit{not B}}$ *Weak*

Example 31 If Spot barks, then Dick will wake up. Dick didn't wake up. So Spot didn't bark.

Analysis This is a valid argument, an example of the indirect way of reasoning with conditionals.

Example 32 If it's the day for the garbageman, then Dick will wake up. It's not the day for the garbageman. So Dick didn't wake up.

Analysis This is weak. Even though the garbageman didn't come, maybe Spot barked or Suzy called early. It overlooks other possible ways the premise could be true and conclusion false.

Example 33 If Maria doesn't call Manuel, then Manuel will miss his class. Maria did call Manuel. So Manuel didn't miss his class.

    Analysis This is weak, denying the antecedent. The "not" in the form indicates a contradictory. Schematically:

If <u>Maria doesn't call Manuel</u>, *then* <u>Manuel will miss his class</u>.
            A   *B*

<u>Maria did call Manuel</u>.   *So* <u>Manuel didn't miss his class</u>.
    not *A*   not *B*

Example 34 If Suzy doesn't call early, then Zoe won't go shopping. Zoe went shopping. So Suzy called early.

    Analysis This is valid, an example of the indirect way. Here the contradictories don't use "not."

Example 35 Zoe won't go shopping if Dick comes home early. Zoe went shopping. So Dick didn't come home early.

    Analysis This is valid, another example of the indirect way.

Example 36 When Johnny comes marching home again, the girls will all laugh and shout. Johnny died in the war. So the girls didn't laugh and shout.

    Analysis This is a weak argument: the girls will laugh and shout anyway—they always do. It's denying the antecedent.

Example 37 If Suzy called early, then Dick woke up. So Dick didn't wake up.

    Analysis The obvious premise to add here is "Suzy didn't call early." But that makes the argument weak, so the argument is unrepairable.

    These invalid forms of arguing are obvious confusions with valid forms, mistakes a good reasoner doesn't make. We'll label them fallacies, and when you see one *don't bother to repair the argument*.

Reasoning in a chain and the slippery slope

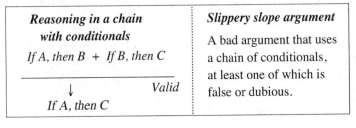

Reasoning in a chain with conditionals	*Slippery slope argument*
If A, then B + *If B, then C* ————————————— ↓ *Valid* *If A, then C*	A bad argument that uses a chain of conditionals, at least one of which is false or dubious.

Example 38 If Dick takes Spot for a walk, then Zoe will cook dinner. And if Zoe cooks dinner, then Dick will do the dishes. So if Dick takes Spot for a walk, then he'll do the dishes. But Dick did take Spot for a walk. So he must have done the dishes.

Analysis This is a valid argument, reasoning in a chain with conditionals followed by the direct way of reasoning with conditionals: we conclude the last consequent because we have the first antecedent.

Example 39 Don't get a credit card! If you do, you'll be tempted to spend money you don't have. Then you'll max out on your card. Then you'll be in real debt. And you'll have to drop out of school to pay your bills. You'll end up a failure in life.

Analysis This is a slippery slope argument, which you can see by rewriting it using conditionals.

Reasoning from hypotheses

One way to try to determine whether a claim is true is to see what follows from it. We make valid or strong arguments starting with the claim as premise. If we can draw a false conclusion, and we've used no other dubious premise, then we can conclude that the hypothesis is false—that's just reducing to the absurd.

Sometimes, though, we get only a new conditional.

Hypotheses and conditionals If you start with an hypothesis *A* and make a good argument for *B*, then you've made a good argument for *If A, then B.*

Example 40 Lee: I'm thinking of majoring in biology.

Maria: That means you'll take summer school. Here's why: You're in your second year now. To finish in four years like you told me you need to, you'll have to take all the upper-division biology courses your last two years. And you can't take any of those until you've finished the three-semester calculus course. So you'll have to take calculus over the summer to finish in four years.

Analysis Maria has not proved that Lee has to go to summer school. Rather, on the assumption (hypothesis) that Lee will major in biology, Lee will have to go to summer school. That is, Maria has proved "If Lee majors in biology, then he'll have to go to summer school."

11 General Claims

"All," "some," "none," and "only"

All means "Every single one, no exceptions."
    Sometimes *all* is meant as
    "Every single one, and there is at least one."

Some means "At least one."
    Sometimes *some* is meant as
    "At least one, but not all."

Which of these readings is best depends on how
the words are used in an argument.

Example 1 All dogs are mammals.
    Analysis This is a true claim.

Example 2 All bank managers are women.
    Analysis This is a false claim on either reading of "all."

Example 3 All polar bears in Antarctica can swim.
    Analysis This is a true claim if you understand "all" as "every
single one." It is false if you understand "all" to include "at least one,"
since there aren't any polar bears in Antarctica.

Example 4 Some dogs bark.
    Analysis This is true on either reading of "some."

Example 5 Some dogs are mammals.
    Analysis This is true if you understand "some" to mean
"at least one." But it is false if you understand "some" to include
"and not all."

There are many different ways to say "all." For example,
the following are equivalent:

    All dogs bark.   Every dog barks.
    Dogs bark.   Everything that's a dog barks.

There are also many ways to say the first reading of "some."
For example, the following are all equivalent:

Some dogs can't bark.	At least one dog can't bark.
There exists a dog that can't bark.	There is a dog that can't bark.

There are also many ways to say that *nothing* or *none* of a collection satisfies some condition. For example, the following are all equivalent:

No dog likes cats.	Nothing that's a dog likes cats.
All dogs do not like cats.	Not even one dog likes cats.

Just as we have to be careful with "only if," we need to be careful with "only."

Example 6   Only bank employees can open the vault at this bank. Pete is a bank employee here. So Pete can open the vault.

Analysis   The argument is weak: Pete might be the janitor. "Only" does not mean "all." "Only bank employees can open the vault" means "Anyone who can open the vault is a bank employee."

> **Only**   *Only S are P* means *All P are S.*

Contradictories of general claims

Example 7   All people want to be rich.
    Contradictory Some people don't want to be rich.

Example 8   Some Russians like chili.
    Contradictory No Russian likes chili.

Example 9   Some women don't want to marry.
    Contradictory All women want to marry.

Example 10   No cat can bark.
    Contradictory Some cat can bark.

Example 11   Every cat hates to swim.
    Contradictory Some cat doesn't hate to swim.

Example 12   Some whales eat fish.
    Contradictory Not even one whale eats fish.

Example 13   Only dogs bark.
    Contradictory Some things that bark are not dogs.

To say that just exactly dogs bark and nothing else, we could say

"Dogs and only dogs bark." The contradictory of that is "Either some dogs don't bark, or some things that bark aren't dogs."

There are many ways to make general claims and many ways to form their contradictories. So at best we have a partial guide.

Claim	Contradictory
All S are P.	Some S is P. Not every S is P.
Some S are P.	No S is P. All S are not P. Not even one S is P.
Some S are not P.	All S are P.
No S is P.	Some S are P.
Only S are P.	Some P are not S. Not every S is P.

Some valid and weak forms of arguments using general claims

The direct way of reasoning with "all"	Arguing backwards with "all"
$\dfrac{\text{All S are P} + a \text{ is S}}{\downarrow} \quad \text{Valid}$ $a \text{ is } P$	$\dfrac{\text{All S are P} + a \text{ is P}}{\downarrow} \quad \text{Weak}$ $a \text{ is } S$

Example 14 All mortgage brokers are honest. Ralph is a mortgage brokers. So Ralph is honest.

    Analysis This is valid, an example of the direct way of reasoning with "all." But though valid, it's not good: the first premise is false, as we learned in the financial crash of 2008.

Example 15 All stockbrokers earn more than $50,000. Earl earns more than $50,000. So Earl is a stockbroker.

    Analysis This is weak, arguing backwards with "all." Earl could be a basketball player or a mortgage broker.

Reasoning in a chain with "all"	**Reasoning in a chain with "some"**
All S are P + All P are Q	Some S are P + Some P are Q
↓ Valid	↓ Weak
All S are Q	Some S are Q

Example 16 Every newspaper the Vice President reads is published by an American publisher. All newspapers published by an American publisher are biased against Muslims. So the Vice President reads only newspapers that are biased against Muslims.

   Analysis This is valid, reasoning in a chain with "all."

Example 17 Some dogs like peanut butter. Some things that like peanut butter are human. So some dogs are human.

   Analysis This is weak, reasoning in a chain with "some."

The direct way of reasoning with "no"	**Arguing backwards with "no"**
All S are P + No Q is P	All S are P + No Q is S
↓ Valid	↓ Weak
No Q is S	No Q is P

Example 18 All corporations are legal entities. No computer is a legal entity. So no computer is a corporation.

   Analysis This is valid, the direct way of reasoning with "no."

Example 19 All nursing students take calculus in their freshman year. No heroin addict is a nursing student. So no heroin addict takes calculus in their freshman year.

   Analysis This is weak, arguing backwards with "no."

Precise generalities and vague generalities

Statistical generalities are easy to evaluate in arguments.

Example 20 72% of all workers at the GM plant say they will vote to strike. Harry works at the GM plant. So Harry will vote to strike.

   Analysis We can say exactly where this argument lands on the strong-weak scale: There's a 28% chance the premises could be true and the conclusion false. That's not good enough to be strong.

Example 21   About 95% of all cat owners have cat-induced allergies. Dr. E's ex-wife has a cat. So Dr. E's ex-wife has cat-induced allergies.
   Analysis   This is a strong argument.

Example 22   Only 4% of all workers on the assembly line at the GM plant didn't get a raise last year. Manuel has worked on the assembly line at the GM plant since last year. So Manuel got a raise.
   Analysis   This is a strong argument, so long as we know nothing more about Manuel.

Most imprecise generalities are too vague to figure in a good argument. For example:

most	a few	a number of	many
a lot	mostly	quite a lot	a bunch of

But ***almost all*** and *a very few* are vague generalities which are clear enough for us to use well in our reasoning.

Example 23   Almost all high school principals have an advanced degree. So the principal at ARF High has an advanced degree.
   Analysis   This is a strong argument. Compare it to the direct way of reasoning with "all."

Example 24   Almost all university professors teach every year. Mary Jane teaches every year. So Mary Jane is a university professor.
   Analysis   This is weak. Mary Jane could be a high school teacher. Compare it to arguing backwards with "all."

Example 25   Almost all dogs like ice cream. Almost all things that like ice cream don't bark. So almost all dogs don't bark.
   Analysis   This is weak. Reasoning in a chain with "almost all" is just as weak as reasoning in a chain with "some."

Example 26   Very few army sergeants tortured prisoners in Iraq. Janet is an army sergeant. So Janet did not torture prisoners in Iraq.
   Analysis   This is a strong argument. Compare it to the direct way of reasoning with "no."

Example 27   All truck drivers have a commercial driver's license. Only a very few beauticians have a commercial driver's license. So only a very few beauticians are truck drivers.
   Analysis   This is strong. Compare it to the direct way of reasoning with "no."

Example 28 All professors get a paycheck at the end of the month. Only a very few people under 25 are professors. So only a very few people under 25 get a paycheck at the end of the month.

Analysis This is weak. Compare arguing backwards with "no."

Examples

Here are some examples that illustrate how to evaluate the strength or validity of an argument that uses general claims. For some you can refer to the forms above. But all of them can be evaluated if you return to the basics: it's not whether the premises and conclusion happen to be true, but whether there is a way for the premises to be true and conclusion false, and if so, how likely that is.

Example 29 Only managers can close out the cash register. George is a manager. So George can close out the cash register.

Analysis This is weak. "Only" does not mean "all." George could be a manager in charge of the stockroom.

Example 30 Everyone who wants to become a manager works hard. The people in Lois' group work hard. So the people in Lois' group want to become managers.

Analysis This is weak. Maybe the workers in Lois' group just want a raise and not the responsibility. The example illustrates a weak form: All *S* are *Q*; all *P* are *Q*; therefore, all *S* are *P*.

Example 31 No taxpayer who cheats is honest. Some dishonest people are found out. So some taxpayers who cheat are found out.

Analysis This is weak. It could be that the only people who are found out are ones who steal.

Example 32 All lions are fierce, but some lions are afraid of dogs. So some dogs aren't afraid of lions.

Analysis This is weak. Maybe all dogs run away before they have a chance to recognize that the lions are afraid of them.

Example 33 Some nursing students aren't good at math. John is a nursing student. So John isn't good at math.

Analysis This is weak. John could be one of the many nursing students who is good at math.

Example 34 Every dog loves its master. Dick has a dog. So Dick is loved.

   Analysis   This is valid, but none of the forms we've studied show that.

Example 35   Almost every dog loves its master. Dick has a dog. So Dick is loved.
   Analysis   This is a strong and good argument.

12 Prescriptive Claims

When we reason, we often want to conclude not only what is, but what ought to be.

Descriptive and prescriptive claims
A claim is *descriptive* if it says what is.
A claim is *prescriptive* if it says what should be.

Every claim is either descriptive or prescriptive. Prescriptive claims are sometimes called *normative*, and descriptive ones are sometimes called *positive*.

Example 1 Drunken drivers kill more people than sober drivers.
Analysis This is a descriptive claim.

Example 2 There should be a law against drunk driving.
Analysis This is a prescriptive claim.

Example 3 Dick: I'm hot.
  Zoe: You should take your sweater off.
Analysis Dick has made a descriptive claim. Zoe responds with a prescriptive claim.

Example 4 The government must not legalize marijuana.
Analysis This is a prescriptive claim, where "must" indicates a stronger idea of "should."

Example 5 The government ought to lower interest rates.
Analysis This is a prescriptive claim.

The words "good," "better," "best," and "bad," "worse," "worst," and other *value judgments* are prescriptive when they carry with them the unstated assumption: "If it's good (preferable, . . .), then we (you) should do it; if it's bad, we (you) should not do it."

Example 6 Drinking and driving is bad.
Analysis This is prescriptive, carrying with it the unstated assumption that we should not do what is bad.

A prescriptive claim either asserts a standard—this is what should be, and there's nothing more fundamental to say than that—or else it assumes another prescriptive claim as standard.

Example 7 Omar: Eating dogs is bad.

Analysis This is a prescriptive claim, since it carries with it the assumption that we should not eat dogs.

Zoe agreed with Omar when he said this to her. But did she really know what standard Omar had in mind? Certainly Omar's claim by itself is not the standard but depends on something more fundamental. Perhaps he's a vegetarian and believes "You should treat all animals humanely, and butchering animals is not humane." Zoe is likely to disagree, since she really enjoys eating a steak.

Or Omar might believe simply "Dogs taste bad." Then he has a standard which requires a further prescriptive one, "You shouldn't eat anything that tastes bad."

Or perhaps Omar believes "Dogs are carnivores, and we shouldn't eat carnivores." That would be a standard which he might support with what he considers a more basic standard, "We should not eat anything forbidden by the standard interpretation of the Koran, and the Koran forbids eating carnivores."

Or perhaps Omar just agrees with what most Americans think, "Dogs should be treated as companions to people and not as food."

Example 8 It's wrong to kill people.

Analysis This is a prescriptive claim. It's usually taken as a standard, rather than assuming another standard.

Example 9 Dr. Wibblitz: The university should stop doing dissection experiments on monkeys.

Analysis This is a prescriptive claim. Suzy agrees because she thinks that monkeys have souls and we shouldn't hurt animals with souls. But Lee disagrees, as he believes that AIDS experiments on monkeys are important. Dr. Wibblitz thinks that such experiments are important, too, yet he thinks that they are too expensive because of the new National Science Foundation regulations. Unless they can agree on a standard by which they mean to judge the claim, they cannot resolve their disagreements.

Debates about prescriptive claims should be about either the standard assumed or whether the claim follows from the standard. We cannot deduce a prescriptive claim from only descriptive claims, for a standard of values is needed first.

> **Is does not imply ought**
> There's no good argument that has a prescriptive conclusion
> with only descriptive premises.

Example 10 Smoking destroys people's health. So we ought to raise
the tax on cigarettes.

Analysis The premise, a descriptive claim, is true. But the conclusion doesn't follow without some prescriptive premise such as "We
should tax activities that are destructive of people's health." The issue
then is why we should believe that.

Example 11 The government should raise the tax rate for the upper
1% of all taxpayers.

Analysis This is a prescriptive claim. Before we can judge
whether to accept it, we need to know what standard lies behind that
"should"—what does the speaker consider a good method of taxation,
and why?

Example 12 ARF High should require students to wear uniforms in
order to minimize gang signs.

Analysis This is really two prescriptive claims: "ARF High
should require students to wear uniforms" and then a reason why,
"We should minimize gang signs (in school)."

Example 13 "I totally don't support prohibiting smoking in bars—
most people who go to bars do smoke and people should be aware that
a bar is a place where a lot of people go to have a drink and smoke.
There are no youth working or attending bars and I just don't believe
you can allow people to go have a beer but not to allow people to have
a cigarette—that's a person's God-given right."

Gordy Hicks, City Councilor, Socorro, NM,
reported in *El Defensor Chieftain*, 7/24/2002

Analysis The implicit standard here for why smoking shouldn't
be prohibited in bars seems to be that society should not establish
sanctions against any activity that doesn't corrupt youth or create harm
to others who can't avoid it. The argument is just as good without the
appeal to God, so by the Guide to Repairing Arguments we can ignore
that. If it turns out that Hicks really does take the standard to be theo-
logical, then the argument he gives isn't adequate.

People who believe that all prescriptive claims are subjective are called *relativists*. They think that all standards—for beauty, morality, and every other value—are relative to what some person or group of people believe. Most people, though, believe that at least some prescriptive claims are objective, such as "You shouldn't torture people."

Often when you challenge someone to make their standard explicit, they'll say, "I just mean it's wrong (right) to me." Yet when you press them, it turns out they're not so happy that you disagree. What they really mean is "I have a right to believe that." Of course they do. But do they have a good *reason* to believe the claim? It's rare that people intend their moral views to be taken as subjective.

Example 14 "The problem with all these criteria is that the choice among them seems entirely arbitrary. [The author cites various conflicting standards on which to base economic policy.] ... I suspect though that the choice of a normative criterion is ultimately a matter of taste."   Stephen Landsburg, *The Armchair Economist*

Analysis This author seems to be a relativist. But he might just be committing the subjectivist fallacy, mistaking lack of agreement for subjectivity.

When a scientist asks us to accept a prescriptive claim, he or she is no longer talking as a scientist but as someone qualified to make value judgments, playing the role of a politician, or philosopher, or priest. *No prescriptive claim follows from any scientific laws or data, for some standard—some value judgment—is required.* It's a bad appeal to authority to accept a prescriptive claim just because a scientist said it.

Example 15 "The science says you've got to reduce emissions of greenhouse gases. The science says you've got to stabilize concentrations of greenhouse gases in the atmosphere. What may be subject to debate is who is to reduce how much."
Rajendra K. Pachauri, Chairman,
UN Intergovernmental Panel on Climate Change

Analysis The science says no such thing. Pachauri is speaking as a politician in making these judgments about where resources should be spent given the consequences of global warming, not as a scientist.

Numbers and Graphs

13 Numbers

We use numbers to measure, to summarize lots of information, and then to compare.

Percentages

Percentages are used to present a summary of numbers. A percentage is a fraction of 100.

percent	fraction	ratio
26%	$26/100$	26 out of 100
92%	$92/100$	92 out of 100
18.1%	$18.1/100$	181 out of 1,000
0.2%	$.2/100$	2 out of 1,000

Sometime percentages greater than 100 are used to indicate an increase.

400%	$400/100$	4 times as much
115%	$115/100$	1.15 times as much

Example 1 52 out of 217 students failed Calculus I last year.

To calculate the percentage of students who failed, take $52/217 =$ 24%, rounded to the nearest percentage.

Example 2 Of 81,173 women tested, 41,829 were allergic to cats.

In percentages, $41,829/81,173 = 51.53\%$ were allergic to cats, rounding to the nearest hundredth of a percent.

Example 3 Last year Ralph's Pet Supply sold 412 dog collars. This year it sold 431.

To calculate the *increase* as a percentage of the previous year's sales, which is the *base*, take the difference and divide by the previous year's number: $(431 - 412)/412 = 4.6\%$ increase in the sale of dog collars, rounding to the nearest tenth of a percent.

Example 4 Last month Piotr Adamowicz's Car Repair took in $59,031 in total receipts. This month it took in $51,287.

To calculate the *decline* in gross receipts as a percentage of last month's sales, which is the base of the comparison, take the difference in gross receipts and divide by the base $(59,031-51,287)/59,031 = 13.1\%$, rounding to the nearest tenth of a percent.

Example 5 Ralph's Pet Supply buys dog collars from a wholesaler for $3.21 and sells them for $6.95.

The store's *markup* is $3.74, which, as a percentage of the price it pays, is $3.74/3.21 = 117\%$, rounding to the nearest percent. Their *cost* as a percentage of what it sells them for is $3.21/6.95 = 46\%$.

Example 6 Last month out of the 47 rats used in Dr. Wibblitz's experiments, 17 died. This month 24 out of 52 rats died. The death rate last month was $17/47 = 36.2\%$ to the nearest tenth of a percent. The death rate this month is $24/52 = 46.2\%$ to the nearest tenth of a percent. The *increase* in the death rate was $(46.2–36.2)/36.2 = 27.6\%$ to the nearest tenth of a percent.

Example 7 Tom sees a stock for $60 and thinks it's a good deal. He buys it; a week later it's at $90, so he sells. He made $30—a 50% gain! His friend Wanda hears about it and buys the stock at $90; a week later it goes down to $60, so she panics and sells the stock. Wanda lost $30—that's a $33^1/_3\%$ loss. The same $30 is a different percentage depending on where you start, that is, depending on the base of the comparison:

$$50\% \uparrow \begin{array}{c} \$90 \\ \$60 \end{array} \downarrow 33^1/_3\%$$

Mean, median, and mode

Another way to present a summary of lots of numbers is by giving the mean, median, or mode of them.

Mean, median, and mode

The **mean** or **average** of a collection of numbers is obtained by: adding the numbers, then dividing by how many items there are.

The **median** is the midway mark: the same number of items above as below.

The **mode** is the number most often attained.

Example 8 For the numbers 7, 9, 37, 22, 109, 9, 11:

The *average* or *mean* is calculated:

Add $7 + 9 + 37 + 22 + 109 + 9 + 11 = 204$
Divide 204 by 7 (the number of items) = 29.14

The *median* is 11.

The *mode* is 9.

Example 9   Dick: It ought to be safe to cross here. I heard that the average depth is only 2 feet.

An average is a useful figure to know only if there isn't
too much variation.

Example 10   The average weight of the children in Ms. Ragini's fifth grade class is 103.7 pounds.
   Analysis   Are most of the children at about that weight, or are there a lot of skinny kids and a few obese ones? More informative would be the median. Better yet, with fewer than 30 children the actual numbers can be given, along with a summary with the median.

Example 11   The median weight of the children in Mr. Humbert's fifth grade class this year is 91 pounds and last year it was 88 pounds.
   Analysis   To allow for a comparison we need to summarize the numbers, which is best done in this case with the medians.

Example 12   The average weight of children in New York's fifth grade classes is 101.72 pounds.
   Analysis   Again, the median would be more informative. But with this large a number of children, the mode could tell us a lot, too.

Example 13   The average wage of concert pianists in the U.S. is less than the average wage of university biology professors.
   Analysis   There's not much variation in the salaries of university biology professors, but there's a huge variation in concert pianists' income ($15,000 to $2,000,000). The modes and medians would allow for a better comparison.

Example 14   Dr. E's final exam
Here are the scores from Dr. E's final exam in critical thinking.

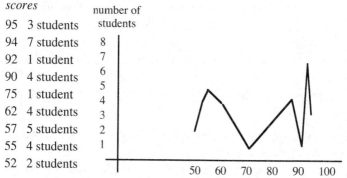

scores	number of students
95	3 students
94	7 students
92	1 student
90	4 students
75	1 student
62	4 students
57	5 students
55	4 students
52	2 students

The grading scale was 90–100 = A, 80–89 = B, 70–79 = C, 60–69 = D, 59 and below = F. When Dr. E's department head asked him how the teaching went, he told her, "Great, just like you wanted, the average mark was 75%, a C." But she knows Dr. E too well to be satisfied. She asks him, "What was the median score?" Again Dr. E replies, "75"—as many got above 75 as below 75. But knowing how clever Dr. E is with numbers, she asks him what the mode score was. Dr. E flushes, "Well, 94." Now she knows something is fishy. When she wanted the average score to be about 75, she was thinking of a graph that looked like:

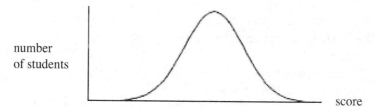

The distribution of the marks should be in a bell-shape: clustered around the median.

Unless you have good reason to believe that the average is pretty close to the median and that the distribution is more or less bell-shaped, the average isn't informative.

Misleading numbers and comparisons

Example 15 There were twice as many rapes as murders in our town last year.

Analysis So? This is a meaningless *apples and oranges* comparison of different kinds of things where there's no common basis for comparing.

Example 16   Prisons are getting worse as breeding grounds for disease in this state. There were 8% more cases of TB among prison inmates this year than last.

Analysis   This is a misleading comparison. If the prison population increased by 16%, then it would be no surprise that the number of cases of TB is going up, though the rate (how many per 1,000 inmates) might be going down.

Example 17   Paid attendance at Learn Your Way Out of Debt seminars is up more than 50% this year!

Analysis   This sounded impressive to Lee until he found that last year 11 people paid for the seminars, and this year 17 did. When the base of a comparison is not given, it's just *two times zero is still zero*.

Example 18   "Identity Theft. Prevention & Repair Kit.
The fastest growing white-collar crime in the US!"
Publication of the Attorney General's Office of New Mexico, 2007

Analysis   So it went from 5 cases to 10? Or from 5,000 to 6,000?

Example 19   *UNM athletes continue to win in the classroom*
"Once again, University of New Mexico student-athletes have made the grade. And then some.

The overall grade-point average for the school's 21 sports was 3.05 in the fall semester, according to the UNM registrar's office. That surpasses the previous best of 3.04, set in the spring of 2003.

It's the eighth time in 12 semesters that the Lobos set a new standard, the figures being released Tuesday."
Mark Smith, *Albuquerque Journal*, 2/23/2005

Analysis   This is a two-times-zero-is-still-zero comparison. The base of the comparison should be the grade-point average for all students, which isn't given. With grade inflation it might be 3.0, so student athletes aren't better than average. Or maybe student athletes were taking easy courses—what's the grade-point average for just the courses they were taking?

Example 20   A report on the radio says unemployment is up 10%.

Analysis   This does not mean unemployment is *at* 10%. It means that if unemployment was 10% it is now 11%, a big increase; or if unemployment was 2%, it's now 2.2%, a very small increase. Unless we know what unemployment was before, the comparison is meaningless.

Sometimes we're given numbers without any context or clear idea of what's being measured.

Example 21 Roadway Congestion
"Cities with highest and lowest roadway congestion index.
A value greater than 1.0 indicates significant congestion.

Highest	Index	Lowest	Index
Los Angeles	1.57	Bakersfield, Calif.	0.68
Washington	1.43	Laredo, Texas	0.73
Miami-Hialeah	1.34	Colorado Springs	0.74
Chicago	1.34	Beaumont, Texas	0.76
San Francisco	1.33	Corpus Christi, Texas	0.78 "

USA Today, 4/13/99

Analysis What are they talking about? What does a "road congestion index" measure? These figures are meaningless to us.

Percentages are meant to summarize lots of numbers. *Percentages without the actual numbers are always misleading.*

Example 22 Last term 22.857% of all Dr. Aloxomani's students failed his organic chemistry class.

Analysis This is a case of *phony precision.* It's accurate but misleading, suggesting that there was a huge sample when it was just 8 out of 25.

Example 23 An article in the journal *Science*, vol. 292, says that mammography screening can reduce the risk of breast cancer fatalities in women ages 50 to 74 by 25%.

Analysis This seems like a real incentive for women that age to get tested. But the article points out that only 2 out of 1,000 women *without symptoms* are likely to die of breast cancer within the next 10 years. So reducing the risk by 25% means that at most 1 more woman in 2,000 who undergoes screening in the next 10 years might be saved.

Sometimes we're asked to believe a claim with numbers when our reaction should be to ask *how could they know those numbers?*

Example 24 Heard on National Public Radio:
"Breast-feeding is up 16% from 1989."

Analysis How could they know that number? Who was looking in all those homes? A survey? Who did they ask? Women chosen randomly? But lots of them don't have infants. Women who visited doctors? But lots of women, lots of poor ones, don't see a doctor. What does "breast-feeding" mean? Does a woman who breast-feeds one day and then stops qualify as someone who breast-feeds? Or one who breast-feeds two weeks? Six months? And up 16% from what base? Maybe NPR is reporting on a reliable survey, but without more information, it's just noise.

Example 25 [From a glossy brochure "Why do I need a water
softener?" by Pentair Water Treatment]
"The Bureau of Statistics found that between 17 and 20.8 cents of every dollar are spent on cleaning products. . . . The bottom line? Soft water can save you thousands of dollars."

Analysis What dollars are they talking about? When you consider the billions of dollars spent by the government on debt and the military—which aren't for cleaning products—you can see that this can't be right. And there's no reason to believe that so much money is spent on cleaning products by individuals. Worse, there's no government agency called "The Bureau of Statistics."

Example 26 "New Mexico Department of Health statistics estimate that of the 115,000 New Mexicans with diabetes, 37,000 don't know they have it."   *El Defensor Chieftain*, Socorro, N.M., 11/9/2005

Analysis If they don't know they have diabetes, how does the Department of Health know? There may be a good way they got this number, but we should suspend judgment about whether it's right unless we're willing to believe everything the Department of Health says.

Even when all the numbers are accurate, and it's all clear, the interpretation of the numbers can be skewed.

Example 27 Harry: Did you hear that new applications for unemployment have fallen since last month and also from this month last year?
Dick: At last the economy is picking up.
Analysis Or there are so many people already out of work that there aren't many left to be fired. *Imagine the possibilities.*

The don't-drop-the-other-shoe technique
Another "statistic" widely quoted in feminist literature comes from the Society for the Advancement of Women's Health. It says that "only 14% (of the National Institutes of Health

clinical trials funding) goes to research 52% of the population." In other words, "women-predominant" diseases, such as breast cancer, get the short end of the stick. Sounds terrible, discriminatory, unfair! But wait a minute. At least 76% of NIH clinical trial grants go to diseases that affect both sexes, such as heart disease and lung cancer. Since 76 + 14 equals 90, whether Washington lobbying groups like it or not, that means that "men-predominant" diseases are getting no more than 10% of the research money while women-predominant diseases get 14%.

It is just such techniques that caused Benjamin Disraeli, 130 years ago, to say that mendacity comes in only three forms, 'lies, damn lies, and statistics.' "

John Steele Gordon, USA Today, 5/21/1999

Socorro
New Mexico
Population 8877
Elevation 4683
Date Est. 1882
Total 15442

14 Graphs

Graphs summarize lots of number claims. When they're done well, as in the next example, they allow for easy, visual comparisons.

Example 1

All of New Mexico Is in Some Stage of Drought.

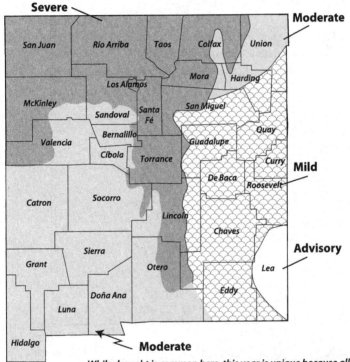

While drought is common here, this year is unique because all of the state is dry and almost one-third is in a severe drought stage

Source: National Resources Conservation Service, http://www.nm.nrcs.usda.gov/drought/drought.htm

Ted Sammis in *Enchantment Magazine*
published by rural electric cooperatives, 2005

But we need to be careful in accepting graphs because they can make—and often conceal—the same kinds of mistakes we've already seen with other comparisons using numbers.

Example 2

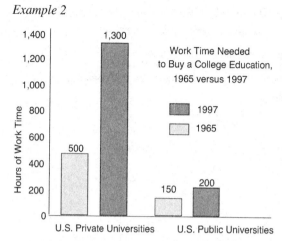

W. Baumol and A. Blinder, *Economics*

Analysis   This graph is from an authoritative source that we're likely to trust, an economics textbook. But it's wrong: the text says that the average hourly wage was about $13 in 1997. So according to the graph the (average?) cost of a college education in 1997 at a U.S. public university was about $13/hour x 200 hours = $2,600. But that wouldn't have been enough for tuition and books in 1997, let alone for four years.

Example 3

Retail sales plunge
Total monthly retail sales:

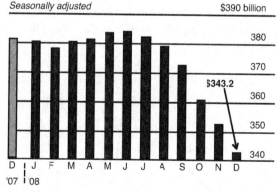

Source: Department of Commerce   Associated Press, 1/15/2009

Analysis   The graph lies. The decrease from the highest sales in July to the lowest sales in December is about 11%, but the height of the bars makes it look like a decrease of 90%.

Whenever a graph doesn't show the baseline—the base of the comparison—it will exaggerate increases and decreases.

Example 4   **Runway incursions on the increase**
Incursions at U.S. airports increased 12 percent
from 2006 to 2007, almost as high as the 2001 peak.

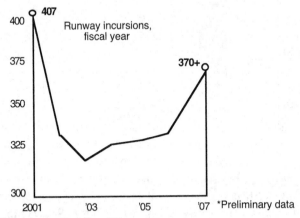

Note: An incursion is any aircraft, vehicle or person
that enters space reserved for takeoff and landing.

Source: Federal Aviation Administration

Associated Press, 12/6/2007

Analysis   The figures are all there, the source is reliable, the graph is easy to read, but a 12% increase in runway incursions is depicted as a 150% increase in the heights of the points.

Example 5   The spacing of the numbers on the axes can affect how we perceive changes that we're comparing. These two graphs present the same data as the previous example.

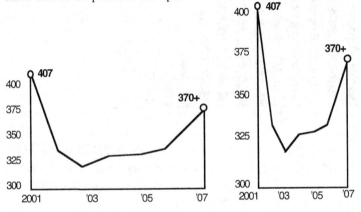

Example 6 An economics text gives this graph:

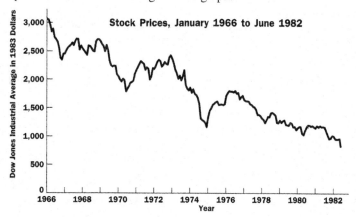

The text remarks that from 1966 to 1982 the prices of stocks were generally going down.

Noting that from 1993 to 1998 stock prices were generally going up, the text then presents the following graph:

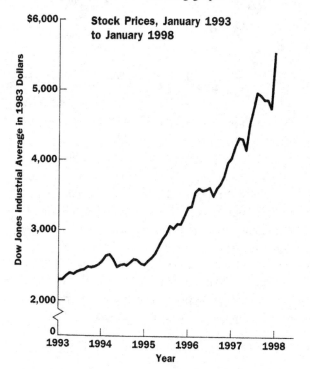

Finally, the text gives a fuller graph.

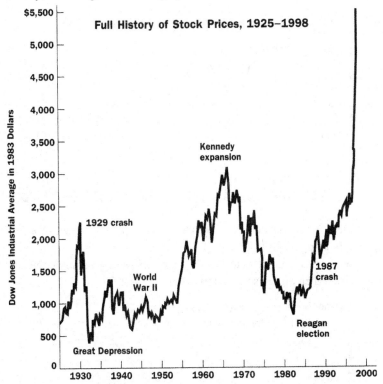

Full History of Stock Prices, 1925–1998

"A much longer and less-biased choice of period (1925–1998) gives a less distorted picture. It indicates that investments in stocks are sometimes profitable and sometimes unprofitable."

W. J. Baumol and A. S. Blinder, Economics

Analysis   Why is the longer period apt for comparison to the present day? If we looked at 1890 onward, we'd have a different picture still (the label "Full History" is wrong). Maybe a better comparison for investing in stocks is with the later periods because of new regulations on buying and selling stocks. The graphs do, however, compensate for inflation by stating the values in 1983 dollars—if they didn't, it would be apples and oranges.

Also note how the steepness of the increases and decreases are exaggerated in the last graph compared to the others.

Example 7

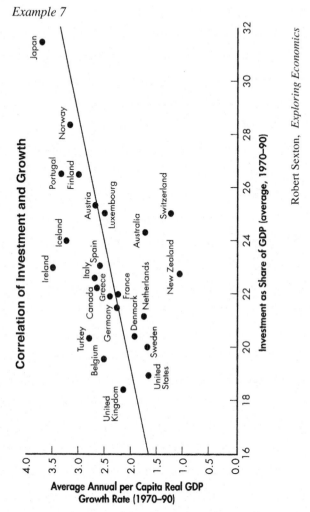

Robert Sexton, *Exploring Economics*

Analysis By drawing the line in this graph, the author is asserting that investment and growth are correlated: both rise together. His premises are the numbers plotted as points. But the picture doesn't obviously support that. Unless you know why and how the author is justified in drawing this line, the conclusion of this implicit argument here is just an appeal to the author's authority.

A graph is a summary of one or more claims or sometimes a whole argument, which we should evaluate by the standards we already have.

Reasoning
about
Experience

15 Analogies

Analogies and comparisons

> **Analogies** A comparison becomes *reasoning by analogy* when it's part of an argument: on one side of the comparison we draw a conclusion, so on the other side we say we should conclude the same.

Example 1 We should legalize marijuana. After all, if we don't, what's the rationale for making alcohol and tobacco legal?
    Analysis Alcohol is legal. Tobacco is legal. Therefore, marijuana should be legal. They are sufficiently similar. This is reasoning by analogy.

Example 2 DDT has been shown to cause cancer in rats. So there's a good chance DDT will cause cancer in humans.
    Analysis This is reasoning by analogy, with an unstated comparison: Rats are like humans. Hence, if rats get cancer from DDT, so will humans.

Example 3 "My love is like a red, red rose." — Robert Burns
    Analysis This is not reasoning by analogy: there's no argument.

    Most reasoning by analogy is typically incomplete, relying on an unstated general principle. Often the value of an analogy is to uncover that principle.

Example 4 Blaming soldiers for war is like blaming firemen for fires. (Background: Country Joe McDonald was a rock star who wrote songs protesting the war in Vietnam. In 1995 he was interviewed on National Public Radio about his motives for working to establish a memorial for Vietnam War soldiers in Berkeley, California, his home and a center of anti-war protests in the '60s and '70s. This claim was his response.)
    Analysis This is a comparison. But it's meant as an argument:

We don't blame firemen for fires.
Firemen and fires are like soldiers and wars.
Therefore, we should not blame soldiers for war.

In what way are firemen and fires like soldiers and wars? They have to be similar enough in some respect for Country Joe's remark to be more than suggestive. We need to pick out important similarities that we can use as premises.

> *Firemen and fires are like soldiers and war*
> wear uniforms
> answer to chain of command
> cannot disobey superior without serious consequences
> fight (fires/wars)
> work done when fire/war is over
> until recently only men
> lives at risk in work
> fire/war kills others
> firemen don't start fires — soldiers don't start wars
> usually like beer

That's stupid: Firemen and soldiers usually like beer. So?

When you ask "So?" you're on the way to deciding if the analogy is good. It's not just any similarity that's important. There must be some crucial, important way that firemen fighting fires is like soldiers fighting wars, some similarity that can account for why we don't blame firemen for fires that also applies to soldiers and war. Some of the similarities listed don't seem to matter. Others we can't use because they trade on an ambiguity, like saying firemen "fight" fires.

We don't have any good guide for how to proceed—that's a weakness of the original argument. But if we are to take Country Joe McDonald's remark seriously, we have to come up with some principle that applies to both sides.

The similarities that seem most important are that both firemen and soldiers are involved in dangerous work, trying to end a problem/disaster they didn't start. We don't want to blame someone for helping to end a disaster that could harm us all.

(‡) Firemen are involved in dangerous work.
     Soldiers are involved in dangerous work.
     The job of a fireman is to end a fire.
     The job of a soldier is to end a war.
     Firemen don't start fires.
     Soldiers don't start wars.

But even with these added to the original argument, we don't get a good argument for the conclusion that we shouldn't blame soldiers for wars. We need a general principle:

> You shouldn't blame someone for helping to end a disaster that could harm others if he didn't start the disaster.

This general principle seems plausible, and it yields a valid argument.

But is the argument good? Are all the premises true? This is the point where the differences between firemen and soldiers might be important.

The first two premises of (‡) are clearly true, and so is the third. But is the job of soldiers to end a war? And do soldiers really not start wars? Look at this difference:

> Without firemen there would still be fires.
> Without soldiers there wouldn't be any wars.

Without soldiers there would still be violence. But without soldiers—any soldiers anywhere—there could be no organized violence of one country against another ("What if they gave a war and nobody came?" was an anti-war slogan of the Vietnam War era).

So? The analogy shouldn't convince. The argument has a dubious premise.

We did not prove that soldiers should be blamed for wars. As always, *when you show an argument is bad you haven't proved the conclusion is false.* You've only shown that you have no more reason than before for believing the conclusion.

Perhaps the premises at (‡) could be modified, using that soldiers are drafted for wars. But that's beyond Country Joe's argument. If he meant something more, then it's his responsibility to flesh it out. Or we could use his comparison as a starting place to decide whether there is a general principle, based on the similarities, for why we shouldn't blame soldiers for war.

Steps in evaluating an analogy
- Is this an argument? What is the conclusion?
- What is the comparison?
- What are the similarities?
- Can we state the similarities as premises and find a general principle that covers the two sides?

- Does the general principle really apply to both sides? Do the differences matter?
- Are the premises really true?
- Is the argument valid or strong?

Examples

Example 5 It's wrong for the government to run a huge deficit, just as it's wrong for any family to overspend its budget.

Analysis The unstated assumption behind this analogy is that what is good for a person or family is also what is good for a country. Without more premises, though, this is unconvincing. There are very big differences between a family and a country: a family doesn't have to repair roads, it can't put up tariffs, nor can it print money.

The *fallacy of composition* is to argue that what is true of the individual is therefore true of the group, or what is true of the group is therefore true of the individual. The differences between a group and an individual are typically too great for such an analogy to be good.

Example 6 "For at least three years in California, about every third teacher hired was brought aboard under an emergency permit, a provisional license that enables people who possess college degrees, but no teaching credentials, to work.

'We wouldn't allow a brain surgeon to learn on the job,' says Day Higuchi, president of the United Teachers Los Angeles, a 41,000-member teachers union. 'Why is it OK to let someone who doesn't know what they're doing teach our kids?' "   *USA Today*, 8/30/1999

Analysis This is an argument, with conclusion (stated as a rhetorical question) that it isn't OK to let someone teach who isn't trained as a teacher. Higuchi, however, needs another premise "If someone doesn't have a teacher's credential, then they don't know what they're doing teaching," which is not so clearly true. The comparison of a brain surgeon with a teacher has too many dissimilarities to be convincing. A teacher saying "Oops" is nothing like a brain surgeon saying "Oops."

Example 7 Suzy: This candy bar is really healthy. Look, on the label it says "All natural ingredients."

Dick: Lard is all natural, too.

Analysis Dick is refuting Suzy's argument that the candy bar
is healthy by using an analogy: the same unstated principle would give
a good argument for lard being healthy, which we know is false.

Example 8 "According to a Food and Drug Administration statement,
'the question of a relationship between brain tumors and aspartame was
initially raised when the agency began considering approval of this
food additive in the mid-1970s.'

However, aspartame was approved for use in 1981.

Since it is an effective insecticide and rodenticide, I can't see any
justification for human consumption."

Ask the Bugman, Richard Fagerlund, *Albuquerque Journal*, 5/9/2009

Analysis This is an analogy from the ill effects of a chemical on
animals and insects to the ill effects on humans. But what are the
similarities between the animals and humans that matter and why don't
the differences matter? We can refute this argument by noting that
chocolate will kill dogs but it's fine (actually great!) for humans.

Example 9 [Concerning the suggestion that the government should
do nothing to rescue the big automakers Chrysler and General Motors
from going bankrupt in 2008.]
"It's easy to demonize the American auto industry. It has behaved with
the foresight of a crack addict for years. But even when people set their
own house on fire, we still dial 9-1-1, hoping to save lives, salvage
what we can and protect the rest of the neighborhood."

Bob Herbert, *The New York Times*, 11/15/2008

Analysis The comparison here is between someone setting his
house on fire and automakers running their business badly. But the
differences seem immense. In particular, fires are immediately
physically dangerous. We shouldn't be convinced by Herbert's
suggestion that the government should rescue the automakers.

Example 10 Downloading computer software from someone you
don't know is like accepting candy from a stranger.

Analysis This becomes an analogy when you consider the
conclusion on one side that your mother probably used to tell you:
you shouldn't accept candy from a stranger. That suggests we should
conclude: you shouldn't download (accept) computer software from
someone you don't know. This looks pretty good, though we need to
come up with a general principle that covers both sides.

Example 11 a. Suppose that good, highly reliable research is announced showing that a liquid derived from eyes removed without anesthetic from healthy cats when applied to human skin reduces wrinkles significantly. Would it be justifiable to do further research and manufacture this oil?

b. Same as (a) except that the liquid is drunk with orange juice and significantly reduces the chance of lung cancer for smokers.

c. Same as (a) except the liquid is mixed with potatoes and eaten and significantly reduces the chance of heart disease and lengthens the lives of women.

d. Same as (a) except that when drunk the liquid kills off all viruses harmful to humans.

Analysis If you said "yes" for some and "no" for others, what differences are there? If you said the same for all, did you reason by analogy? What general principle did you use? Would it apply if we replaced "cats" with "dogs"? Evaluating the analogies and disanalogies can lead to insight about our ethical assumptions.

Analogies and the law

Most analogies are not made explicit enough to serve as good arguments. But in the law, analogies are presented as detailed, carefully analyzed arguments, with the important similarities pointed out and a general principle stated.

Laws are often vague, or situations come up which no one ever imagined might be covered by the law: do the tax laws for mail-order purchases apply to the Internet? Similarities or differences have to be pointed out, general principles enunciated. Then those principles have to be respected by other judges. That's the idea of precedent or common law.

> "The basic pattern of legal reasoning is reasoning by example. It is reasoning from case to case. It is a three-step process described by the doctrine of precedent in which a proposition descriptive of the first case is made into a rule of law and then applied to a next similar situation. The steps are these: similarity is seen between cases; next the rule of law inherent in the first case is announced; then the rule of law is made applicable to the second case."
>
> Edward H. Levi, *An Introduction to Legal Reasoning*

But why should a judge respect how earlier judges ruled? Those decisions aren't actually laws.

Imagine getting thrown in jail for doing something that's always been legal, yet the law hasn't changed. Imagine running a business and suddenly finding that something you did, which before had been ruled safe and legal in the courts, now leaves you open to huge civil suits because a judge decided differently this week. If we are to live in a society governed by laws, the law must be applied consistently. It's rare that a judge can say that past decisions were wrong.

Only a few times has the Supreme Court said that all rulings on one issue, including rulings the Supreme Court made, are completely wrong. Brown vs. the Board of Education said that segregation in schools, which had been ruled legal for nearly a hundred years, is now illegal. Roe vs. Wade said that having an abortion, which had been ruled illegal for more than a century, is now legal. Such decisions are rare. They have to be. They create immense turmoil in the ways we live. We have to rethink a lot. And we can't do that regularly.

So what does a judge do when she's confronted by fifteen cases that were decided one way, the case before her falls under the general principle that was stated to cover those cases, yet her sense of justice demands that she decide this case the other way? She looks for differences between this case and those fifteen others. She tweaks the general principle just enough to get another principle that covers all those fifteen cases, but doesn't include the one she's deciding. She makes a new decision that now must be respected or overthrown.

Example 12 The Supreme Court has decided it is a constitutional right for a doctor to terminate medical treatment that prolongs the life of a terminally ill or brain-dead person, so long as the doctor acts according to the wishes of that person (*Cruzan vs. Director, Missouri Department of Health*, 497 U.S. 261). So the Supreme Court should decide that assisting someone to commit suicide, a person who is terminally ill or in great suffering, as Dr. Kevorkian did, is a constitutionally protected right (*Compassion in Dying vs. State of Washington*).

Analysis The court can decide narrowly, by saying this new case is not sufficiently like *Cruzan*, or broadly, by enunciating a principle that applies in both cases or else distinguishes between them. Or it can bring in more cases for comparison to try to decide what general principle applies. (You can look up on the Internet how the court decided.)

16 Generalizing

> **Generalizations** A *generalization* is an argument in which we conclude a claim about a group, called the **population**, from a claim about some part of it, the **sample**. Sometimes we call the general claim which is the conclusion the generalization. Plausible premises about the sample are called the **inductive evidence** for the generalization.

The population and the sample

Example 1 In a study of 5,000 people who owned pets in Anchorage, Alaska, dog owners expressed higher satisfaction with their pets and with their own lives. So dog owners are more satisfied with their pets and their own lives than other pet owners.

Analysis *Sample*: The 5,000 people who were surveyed in Anchorage.

Population: Pet owners everywhere.

Example 2 Of potential customers surveyed, 72% said that they liked "very much" the new green color that Yoda plans to use for its cars. So about 72% of all potential customers will like it.

Analysis *Sample*: The group of potential customers who were interviewed.

Population: All potential customers.

This is a statistical generalization.

Example 3 Every time the minimum wage is raised, there's squawking that it will cause inflation and increase unemployment. And every time it doesn't. So watch for the same bad arguments again this time.

Analysis The unstated conclusion is that raising the minimum wage will not cause inflation or increase unemployment. This is a generalization from the past to the future.

Sample: All times in the past that the minimum wage was raised.

Population: All times it was raised or will rise.

Example 4 The doctor tells you to fast from 10 p.m. At 10 a.m. she gives you glucose to drink. Forty-five minutes later she takes some of your blood and has it analyzed. She concludes you don't have diabetes.

Analysis   Sample: The blood the doctor took.
Population: All the blood in your body.

Example 5   Maria goes to the city council meeting with a petition signed by all the people who live on her block requesting that a street light be put in. Addressing the city council, she says, "Everyone on this block wants a street light here."

Analysis   This is not a generalization. There's no argument from some to more: the sample equals the population.

Representative samples

What constitutes a good generalization?   If we have a sample that's just like the entire collection, we can trust a generalization we make from it.

> **Representative sample**   A sample is *representative* if no one sub-group of the whole population is represented more than its proportion in the population. A sample is **biased** if it is not representative.

How can we get a representative sample?   You might think we can get one by making sure we have no intentional bias when we choose it. That's called **haphazard sampling**, and it's not reliable.

Example 6   For his sociology class, Tom decided he'd try to determine the attitudes of students about sex before marriage by giving a questionnaire to the first 20 students he met coming out of the student union.

Analysis   Tom had no intentional bias in choosing his sample. But there's no reason to believe it's representative. Those students might all be coming from a meeting of the Green Party, or the student Bible Society, or . . . .

Example 7   Zoe reckons she can do better. She enlists three of her friends to hand out the questionnaire to the first 20 students they meet coming out of the student union, the administration offices, and the largest classroom building at 9 a.m., 1 p.m., and 6 p.m.

Analysis   Trying harder to get rid of intentional bias won't guarantee that a sample is representative: perhaps all the players on the intercollegiate sports teams were gone for the day. They need a better way to choose a sample.

> ***Random sampling***   A sample is *chosen randomly* if at every choice there is an equal chance for any of the remaining members of the population to be picked.

Example 8   Tom assigns a number to each student listed in the student directory, writes the numbers on slips of paper, puts them in a fishbowl, and has Suzy draw out one number at a time. Then he gives his questionnaire to the students whose numbers are drawn.

   Analysis   Probably this would be a random selection. A simpler way to get a random selection is to use prepared tables of random numbers which can be generated by a spreadsheet program for a home computer. For Tom's survey he can take the list of students from the directory, and if the first number on the table is 413, he would pick the 413th student on the list; if the second number is 711, he'd pick the 711th student on the list; and so on, until he has a sample that's big enough.

   Random sampling is very likely going to yield a sample that's close to being representative. That's because of the ***law of large numbers***, which says, roughly, that if the probability of something occurring is X percent, then over the long run the percentage of times that happens will be about X percent.

Example 9   The probability that a fair coin when flipped will land heads is 50%. So though you may get a run of 8 tails, then 5 heads, then 4 tails, then 36 heads to start, in the long run, repeating the flipping, eventually the number of heads will tend toward 50%.

Example 10   Of the 20,000 students at Tom's school, 500 are gay males. So the chance that one student picked at random would be a gay male is: 500/20,000 = 1/40. If Tom were to pick 300 students at random, the chance that half of them would be gay is very, very small. It's very likely, however, that 7 or 8 (about 1/40 of 300) will be gay males.

Example 11

Analysis   Dick is confused about the law of large numbers. The ball could land on red 100 times in a row, and black could even out by coming up just one more time than red every 100 spins for the next 10,000 spins. The **gambler's fallacy** is to reason that a run of events of a certain kind makes a run of contrary events more likely to even up the probabilities. The long run can be very long indeed.

"In the long run we're all dead." — John Maynard Keynes

If we choose a large sample randomly, the chance is very high that it will be representative, since the chance of any one subgroup being over-represented is small — not nonexistent, but small. It doesn't matter if we know anything about the composition of the population in advance. After all, to know how many homosexuals there are, and how many married women, and how many men, and . . . you'd need to know almost everything about the population in advance. But that's what we do samplings to find out.

With a random sample we have good reason to believe the sample is representative. A sample chosen haphazardly might give a representative sample — but we have no good reason to believe it will.

Weak Argument	*Strong Argument*
Sample is chosen *haphazardly*. Therefore, the sample is representative.	Sample is chosen *randomly*. Therefore, the sample is representative.
Lots of ways the sample could be biased.	Very unlikely the sample is biased if it's large enough.

Example 12   The classic example that haphazard sampling can be bad even with an enormous sample is the poll done in 1936 by *Literary Digest*. The magazine mailed out 10,000,000 ballots asking who the person would vote for in the 1936 presidential election. It received 2,300,000 back. With that huge sample, the magazine confidently predicted that Alf Landon would win. Roosevelt received 60% of the votes, one of the biggest wins ever. What went wrong? The magazine selected its sample from lists of its own subscribers and telephone and automobile owners. In 1936 that was the wealthy class, which preferred Alf Landon.

Example 13 "We recruited participants at six busy locations in Zurich, Switzerland. Eligible participants were randomly approached and asked whether they would agree to take part in the study. We approached 272 pedestrians, and 185 (68%) were willing to take part. . . .

In this sample, Swiss citizens did not know more than a third of MMK [minimum medical knowledge]. We found little improvement from this low level within groups with medical experience (personal or professional), suggesting that there is a consistent and dramatic lack of knowledge in the general public about the typical signs of and risk factors for important clinical conditions."

"Do citizens have a minimum medical knowledge? A survey"
L. Bachmann, F. Gutzwiller, M. Puhan, J. Steurer, C. Steurer-Stey,
and G. Gigerenzer, *BMC Medicine*, vol. 5, no. 14, 2007

Analysis This is just haphazard sampling and there's no reason to believe that the people interviewed are representative of all Swiss, much less "the general public." Yet this was published in a peer-reviewed journal, albeit one that's only online and has an ad for "Science Singles" at the top of its homepage.

Criteria for a good generalization

Having a sample that we've good reason to think is close to representative is not enough for a generalization to be good. We also need that the sample is big enough. How big? Roughly, the idea is to measure how much more likely it is that the generalization is going to be accurate as we increase the number in our sample.

Example 14 If Lee wants to find out how many people in his class of 300 biology students are spending 10 hours a week on the homework, he might ask 15 or 20. If he interviews 30, he might get a better picture. But there's a limit. After he's asked 100, he probably won't get a much different result if he were to ask 150. And if he's asked 200, it's not likely his generalization will be different if he asks 250.

Example 15 Of the 20,000 students at Tom's school, 500 are gay males. If Tom chooses randomly just eight students to interview, one might be gay, and from that tiny sample he might mistakenly infer that 12% of the students at his school are gay males.

Example 16 The makers of Doakes toothpaste proudly announce that users reported a 25% reduction in cavities, as certified by an independent laboratory.

Analysis This sounds impressive until you look up the study and find that the researchers followed only 12 people. With such a small sample, by chance they might find that the six people who used the toothpaste got 3 cavities and the other six got 4.

Generalizing from a sample that's obviously too small is called a *hasty generalization* based on *anecdotal evidence*. Often we can rely on common sense to evaluate whether a sample is big enough. But when we generalize to a very large population, say 2,500, or 25,000, or 250,000,000, how big the sample should be cannot be explained without at least a mini-course on statistics. In evaluating statistical generalizations, we have to expect that the people doing the sampling have looked at enough examples, which is reasonable if it's a respected organization or a well-known polling company. Surprisingly, 1,500 is typically adequate for a sample size when surveying all adults in the United States.

How big the sample needs to be depends also on how much *variation* there is in the population regarding the aspect you're investigating. If we know in advance that there is very little variation, then a small sample chosen haphazardly will do. When there's lots of variation or you don't know how much variation there is, you need a large sample, and random sampling is the best way to get that.

Example 17 It's incredible how much information they can put on a CD. I just bought one that contains five movies.

Analysis This is a good generalization. The unstated conclusion is that every CD can contain as much information as this one that has the movies on it. There is little variation in the production of CDs for computers, so a sample of one is sufficient.

A large representative sample can still lead to a bad generalization if the sample isn't studied well.

Example 18 The doctor taking your blood to see if you have diabetes won't get a reliable result if her test tube isn't clean or she forgets to tell you to fast the night before. You won't find out the real attitudes of students about tuition if you ask a biased question. Picking a random sample of bolts won't help you determine whether the bolts are OK if all you do is inspect them by eye and not with a microscope or a stress test.

Example 19 Surveys on sexual habits are notorious for being inaccurate: invariably women report that the number of times they engaged in sexual intercourse with a man in the last week, or month, or year is much lower than the reports that men give of sexual intercourse with a woman during that time. The figures are so different that it would be impossible for both groups to be answering accurately.

Generally, questionnaires and surveys are problematic because questions need to be formulated without bias and the interviewer has to rely on the respondents answering truthfully.

Example 20 One of the questions on the "Official 2015 Democratic Party Survey" was:

"Do you support his [President Obama's] plan to close other tax loopholes to simplify the tax code so that corporations and the ultra-wealthy will pay their fair share?"

Analysis There's no reason to believe any generalization from such a biased question. This is propaganda masquerading as a survey.

Example 21 Maria asked all but three of the 36 people in her class whether they've ever used cocaine. Only two said "yes." So Maria concludes that almost no one in the class has used cocaine.

Analysis This is a bad generalization. The sample is big enough and probably representative, but it's not studied well. People are not likely to admit to a stranger that they've used cocaine. An anonymous questionnaire is needed.

Example 22 "More than four million people younger than 21 drove under the influence of drugs or alcohol last year, according to a government report released Wednesday. That's one in five of all Americans aged 16 to 20."   Associated Press, 12/30/2004

Analysis We don't know whether they used an anonymous questionnaire, so we have no reason to accept their generalization.

Premises needed for a good generalization
- The sample is representative.
- The sample is big enough.
- The sample is studied well.

The margin of error and the confidence level

It's never reasonable to believe a statistical generalization whose conclusion is too precise.

Example 23   In a survey, 27% of the people in Nantucket who were interviewed said they wear glasses, so 27% of all people in Nantucket wear glasses.

Analysis   No matter how many people in Nantucket are surveyed, short of all of them, we can't be confident that exactly 27% of all people in the town wear glasses. Rather, "27%, more or less, wear glasses" is the right conclusion.

That "more or less" can be made fairly precise according to a theory of statistics. The ***margin of error*** gives the range in which the actual number for the population is likely to fall. The ***confidence level*** measures how strong the argument is for the statistical conclusion, where the survey method and responses are taken as premises.

Example 24   The opinion poll says that when voters were asked their preference, the incumbent was favored by 53% and the challenger by 47%, with a margin of error of 2% and a confidence level of 95%. So the incumbent will win tomorrow.

Analysis   From this survey they're concluding that the percentage of all voters who favor the incumbent is between 51% and 55%, while the challenger is favored by between 45% and 49%. "The confidence level is 95%" means that there's a 95% chance it's true that the actual percentage of voters who prefer the incumbent is between 51% and 55%. If the confidence level were 70%, then the survey wouldn't be reliable: there would be a 3-out-of-10 chance that the conclusion is false. Typically, if the confidence level is below 95%, results won't be announced.

The bigger the sample, the higher the confidence level and the lower the margin of error. The problem always is to decide how much it's worth in extra time and expense to increase the sample size in order to get a stronger argument.

Example 25   With a shipment of 30 insulating tiles, inspecting 3 and finding them OK would normally allow you to conclude that all the tiles are OK. But if they're for the space shuttle, where a bad tile could doom the spacecraft, you'd want to inspect each and every one of them.

Risk doesn't change how strong an argument we have, only how strong an argument we want before we'll accept the conclusion.

Examples

Example 26 Flo: Every time I've seen a stranger at Dick's gate, Spot has barked. So Spot will always bark at strangers at Dick's gate.
Analysis This is a bad generalization. The sample is chosen haphazardly, so there's no reason to believe it's representative.

Example 27 Dick: Why does the phone always ring when I'm in the shower?
Analysis **Selective attention** is a common mistake in generalizing: we note only what's unusual and forget the other times.

Example 28 In a test of 5,000 cattle from Manitoba, none of them were found to be infected with mad cow disease. So it's pretty likely that no cattle in Canada have mad cow disease.
Analysis This is a bad generalization. There's no reason to think that the sample is representative. At best the evidence could lead to a conclusion about all cattle in Manitoba.

Example 29 Suzy: My grandmother was diagnosed with cancer seven years ago. She's refused any treatment that was offered to her over the years. She's perfectly healthy now and doing great. The treatments for cancer are just a scam to get people's money.
Analysis This is a bad generalization from anecdotal evidence.

Example 30 Dick: A study I read said people with large hands are better at math.
Suzy: I guess that explains why I can't divide.
Analysis Perhaps the study was done carefully with a random sample. But you don't need a study to know that people with large hands do better at math: babies have small hands, and they can't even add. "All people" is the wrong population to study.

Example 31 According to the National Pork Producers Council (www.nppc.org), average hog market weight is 250 pounds, and it takes about 3.5 pounds of feed to produce 1 pound of live hog weight.
Analysis It's a good appeal to authority to accept this generalization given by the National Pork Producers Council. Though we don't have access to the data it used or it was obtained, it's a big enough

organization to employ good statisticians, and it has no reason to lie
to its own members.

Example 32 Lee: Every rich person I've met invested heavily in the
stock market. So I'll invest in the stock market, too.

Analysis This is a confused attempt to generalize. Perhaps Lee
thinks that the evidence he cites gives the conclusion that if you invest
in the stock market, you'll get rich(er). But that's arguing backwards,
confusing (1) "If you invest in the stock market, you'll get rich" with
(2) "If you're rich, then you invested in the stock market." The population for (1) is all investors in the stock market, not just the rich ones.
It's a case of selective attention.

Example 33 Suzy: I've been studying this astrology book seriously.
I think you should definitely go into science.

    Lee: I've been thinking of that, but what's astrology got to do
           with it?

    Suzy: Your birthday is in late January, so you're an Aquarian?

    Lee: Yeah, January 28.

    Suzy: Well, Aquarians are generally scientific but eccentric.

    Lee: C'mon. That can't be right.

    Suzy: Sure it is. Copernicus, Galileo, and Thomas Edison were
           all Aquarians.

Analysis This is a bad generalization based on anecdotal evidence.
It's just selective attention, reasoning backwards.

Example 34 A "Quality of Education Survey" was done by the Las
Cruces, New Mexico school district in 2012. Forms were sent out to
the parents of 25,000 students, and the public was invited to fill out
forms online. The numbers of respondents were:

Parents	6,448
No response	2,380
Teacher at school	303
Guardian of student	300
Staff at school	176
Other relative of student	64
Other interested party	39
Not known	31

The responses were grouped together. For one of the questions the
results were:

"The school staff maintains consistent discipline that is
conducive to learning"

Strongly agree	30.71%
Agree	54.17%
Disagree	7.16%
Strongly disagree	2.88%
No opinion	3.58%
No response	1.47%

Analysis There's no reason to think that the sample is representative, despite the large number of responses. There's no reason to think that those who responded know what goes on in the school. Yet the Department of Education requires that the school district adjust its programs and methods to improve the "scores" on this survey.

Example 35 Of the chimpanzees fed one-quarter pound of chocolate per day in addition to their usual diet, 72% became obese within two months. Therefore, it is likely that most humans who eat one-quarter pound of chocolate per day in addition to their usual diet will become obese within two months.

Analysis A generalization is needed to make this analogy good: 72% of all chimpanzees, more or less, will become obese if fed one-quarter pound of chocolate per day in addition to their usual diet. Whether this is a good generalization will depend on whether the researchers can claim that their sample is representative. The analogy then needs a claim about the similarity of chimpanzee physiology to human physiology in order to be a good argument. Trying to formulate such a claim will make it clear that it isn't the same quantity of chocolate but the amount relative to the body weight of the chimpanzees and the humans that should be the same.

17   Cause and Effect

Describing causes and effects

What exactly is a cause?

Example 1   Last night Dick said:

> Spot made me wake up.

Spot caused Dick to wake up. But it's not just that Spot existed.
It's what he was doing that caused Dick to wake up:

Spot's barking
caused Dick
to wake up.

So Spot's barking is the cause? What kind of thing is that?
The easiest way to describe the cause is to say:

> Spot barked.

The easiest way to describe the effect is to say:

> Dick woke up.

Causes and effects can be described with claims.

Causal claims   A *causal claim* is a claim that is or can be
rewritten as *X causes (caused) Y*.
  A *particular* causal claim is one in which a single claim can
describe the (purported) cause, and a single claim can describe the
(purported) effect. A *general* causal claim is a causal claim that
generalizes many particular causal claims.

Example 2   Spot caused Dick to wake up.
  Analysis   This is a particular causal claim, where the purported
cause can be described by the single claim "Spot was barking" and the

purported effect by "Dick woke up." We might generalize from this particular cause and effect to the claim "Very loud barking by a dog near someone when he is sleeping *causes* him to wake up, if he's not deaf." That is a general causal claim. For it to be true, lots of particular causal claims have to be true.

Example 3 The police car's siren got Dick to pull over.
    Analysis This is a particular causal claim. The purported cause can be described by "The police car had its siren going," and the purported effect by "Dick pulled over."

Example 4 The speeding ticket Dick got made his auto insurance rate go up.
    Analysis This is a particular causal claim. The purported cause is "Dick got a speeding ticket," and the purported effect is "Dick's auto insurance went up."

Example 5 Speeding tickets make people's auto insurance rates go up.
    Analysis This is a general causal claim. For it to be true all particular causal claims as in the previous example have to be true.

Example 6 Penicillin prevents serious infection.
    Analysis What is the cause? The existence of penicillin? No, it's that penicillin is administered to people in certain amounts at certain stages of their infections. What's a "serious infection"? This is too vague to count as a causal claim.

Example 7 Lack of rain caused the crops to fail.
    Analysis The purported cause here is "There was no rain," and the purported effect is "The crops failed." This example was true a few years ago in the Midwest. A cause need not be something active; almost any claim that describes the world could describe a cause.

Necessary conditions for cause and effect
What conditions are needed for a causal claim to be true?

The cause and effect both happened.
That is, the claim describing the cause and the claim describing the effect are both true. We wouldn't say that Spot's barking caused Dick to wake up if Spot didn't bark or Dick didn't wake up.

***It's (nearly) impossible for the cause to happen and the effect not
to happen.***
It has to be (nearly) impossible for the claim describing the cause to
be true and the claim describing the effect to be false. It can't be just
coincidence that Dick woke up when Spot barked.

This condition is the relation of premises to conclusion in a valid or
strong argument. But here we're not trying to convince anyone that the
conclusion is true: we know that Dick woke up. What we can carry
over from our study of arguments is how to look for possibilities—
ways the premises could be true and the conclusion false—to determine
if there is cause and effect. As with arguments we will often need to
supply unstated premises to show that the effect follows from the cause.

Example 8   A lot has to be true for it to be impossible for "Spot
barked" to be true and "Dick woke up" to be false:

Dick was sleeping soundly up to the time that Spot barked.
Spot barked at 3 a.m.
Spot was close to where Dick was sleeping.

We could go on forever. But as with arguments, we state what we
think is important and leave out the obvious. If someone challenged
us, we could add "There was no earthquake at the time"—but we just
assume that as part of the normal conditions.

Normal conditions   The *normal conditions* for a causal claim
are the obvious and plausible claims that are needed to establish
that the relation between the purported cause and the purported
effect is valid or strong.

Example 9   Very loud barking by a dog that's near someone who is
sleeping *causes* him to wake up, if he's not deaf.

Analysis   For a general causal claim like this one, the normal
conditions won't be specific just to the one time Spot woke Dick,
but will be general.

The cause precedes the effect.
We wouldn't accept that Spot's barking caused Dick to wake up if Spot
began barking only after Dick woke up. The cause has to precede the
effect. That is, "Spot barked" became true before "Dick woke up"
became true.

The cause makes a difference.
If there were no cause, there would be no effect.

Example 10 Dr. E has a desperate fear of elephants. So he buys a special wind chime and puts it outside his door to keep the elephants away. He lives in Cedar City, Utah, at 6,000 feet above sea level in a desert, and he confidently claims that the wind chime causes the elephants to stay away. After all, ever since he put up the wind chime he hasn't seen any elephants.

Analysis Why are we sure the wind chime being up did not cause elephants to stay away? Because even if there had been no wind chime, the elephants would have stayed away. Which elephants? All elephants. The wind chime works, but so would anything else. The wind chime doesn't make a difference.

There is no common cause.
We don't say that night causes day because there's a common cause of both "It was night" and "It is now day," namely, "The Earth is rotating relative to the sun."

Example 11 Dick: Zoe is irritable because she can't sleep properly.

Tom: Maybe it's because she's been drinking so much espresso that she's irritable and can't sleep properly.

Analysis Tom hasn't shown that Dick's causal claim is false by raising the possibility of a common cause. But he does put Dick's claim in doubt. We have to check the other conditions for cause and effect to see which causal claim seems most likely.

In summary, here are the conditions required for a causal claim to be true when causes and effects are described by claims.

Necessary conditions for cause and effect

- The cause and effect both happened (both claims are true).
- It's (nearly) impossible for the cause to happen (be true) and the effect not to happen (be false), given the normal conditions.
- The cause precedes the effect.
- The cause makes a difference—if the cause had not happened, the effect would not have happened, given the normal conditions.
- There is no common cause.

These are necessary conditions. In practice, however, we treat them as sufficient, too. Here's an example that shows how to check all these steps in evaluating a causal claim.

Example 12   The cat made Spot run away.

Cause   What is the cause? It's not just the cat. How can we describe it with a claim? Perhaps "A cat meowed close to Spot."

Effect   Spot ran away.

Cause and effect each happened   The effect is clearly true. The cause is highly plausible: almost all things that meow are cats.

Cause precedes effect   Yes.

It's (nearly) impossible for the cause to be true and effect false
What needs to be assumed as "normal" here? Spot is on a walk with Dick. Dick is holding the leash loosely enough for Spot to get away. Spot chases cats. Spot heard the cat meow. We could go on, but this seems enough to guarantee that it's unlikely that the cat could meow near Spot and Spot not chase it.

The cause makes a difference   Would Spot have run away even if the cat had not meowed near him? Apparently not, given those normal conditions, since Dick seems surprised that Spot ran off. Perhaps he would have, though, even if he'd only seen the cat. But that apparently wasn't the case. So let's revise the cause to be "Spot wasn't aware a cat was near him, and the cat meowed close to Spot." Now we can reasonably believe that the cause made a difference.

Is there a common cause?   Perhaps the cat was hit by a meat truck and lots of meat fell out, and Spot ran away for that? No, Spot wouldn't have barked. Nor would he have growled.

Perhaps the cat is a hapless bystander in a fight between dogs, one of which is Spot's friend. We do not know if this is the case. So it is possible that there is a common cause, but it seems unlikely.

Evaluation We have good reason to believe the original claim on the revised interpretation that the cause is "Spot wasn't aware a cat was near him, and the cat meowed close to Spot, and Spot heard it."

These are the steps we should go through to establish a causal claim. If we can show that one of them fails, though, there's no need to check all the others.

Common mistakes in reasoning about cause and effect

Tracing the cause too far back in time
It's sometimes said that the cause must be close in space and time to the effect. But the astronomer is right when she says that a star shining caused the image on the photograph, even though that star is very far away and the light took millions of years to arrive. The problem isn't how distant in time and space the cause is from the effect. The problem is how much has come between the cause and effect—whether we can specify the normal conditions. *When we trace a cause too far back, the problem is that the normal conditions begin to multiply.* There are too many conditions for us to imagine what would be needed to establish that it's impossible for the cause to have been true and the effect false. When you get that far, you know you've gone too far.

Example 13 My mother missed the sign-up to get me into Kernberger Preschool, and that's why I've never been able to get a good job.
    Analysis This is tracing the cause too far back.

Reversing cause and effect
If reversing cause and effect sounds just as plausible as the original claim, investigate the evidence further before making a judgment.

Example 14 Suzy: Sitting too close to the TV ruins your eyesight.
    Zoe: How do you know?
    Suzy: Well, four of my grade-school friends used to sit really close
        to the TV, and all of them wear really thick glasses now.
    Zoe: Maybe they sat so close because they had bad eyesight.
    Analysis Zoe hasn't shown that Suzy's claim is false. But her suggestion that cause and effect are reversed raises sufficient doubt not to accept Suzy's claim without more evidence.

Looking too hard for a cause

We look for causes because we want to understand, so we can control our future. But sometimes the best we can say is that it's *coincidence*.

Example 15 Before your jaw drops open in amazement when a friend tells you a piano fell on her old piano teacher the day after she dreamt she saw him in a recital, remember the law of large numbers: If it's possible, given long enough, it'll happen. After all, most of us dream, say, one dream a night for at least 50 million adults in the U.S. That's 350 million dreams per week. With the elasticity in interpreting dreams and what counts as a "dream coming true," it would be amazing if a lot of dreams didn't "accurately predict the future."

But doesn't everything have a cause? Shouldn't we look for it? For much that happens in our lives we won't be able to figure out the cause—we just don't know enough. We must, normally, ascribe lots of happenings to chance, to coincidence, or else we have paranoia and end up paying a lot of money to phone psychics.

Example 16 A woman in New York cuts her hand with a sharp knife. At just that moment her mother in Montana 2,500 miles away feels a pain in the same hand. Coincidence?
  Analysis Yes! That's what coincidence is.

Post hoc ergo propter hoc ("after this, therefore because of this")

It's a mistake to argue that there is cause and effect just because one claim became true after another.

Example 17 Lee: I scored well on that last exam and I was wearing my red-striped shirt. I'd better wear it every time I take an exam.
  Analysis This is *post hoc* reasoning.

Example 18 A recent study showed that everyone who uses heroin started with marijuana. So smoking marijuana causes heroin use.
  Analysis And they probably all drank milk first, too. Without further evidence this is just *post hoc* reasoning.

Claiming that a correlation by itself establishes cause and effect is the *correlation-causation fallacy*. It's just a pumped-up version of *post hoc* reasoning or reversing cause and effect.

The best way to avoid making common mistakes in reasoning about cause and effect is to experiment. Conjecture possible causes, then by experiment eliminate them until there's only one. Check that one: Does it make a difference? If the purported cause is eliminated, is there still the effect? Often we can't do an experiment, but we can do an imaginary one. That's what we always do in reasoning well: *Imagine the possibilities*.

Examples

Example 19 The President's speech on farm issues made the price of corn rise 17% the next day.

Analysis The purported cause is "The President gave a speech on farm issues," and the purported effect is "The price of corn rose 17% the next day." Did the purported cause make a difference? A few hours after the President's speech, crop reports were released that showed the corn harvest would be down 9% due to drought. Those reports alone would have been enough to ensure higher corn prices. We have no reason to believe that the President's speech was the cause.

Example 20 "Disappointing job creation, Hungary woes send markets reeling"   Headline, Associated Press, 6/5/2010

Analysis Every day newswriters pick out what they consider the most prominent piece of good news if the market went up, or bad news if the market went down, and ascribe the change in the stock market to that. That's just *post hoc* reasoning.

Example 21 Flo: Salad makes you fat. I know 'cause Wanda's really fat and she's always eating salad.

Analysis Flo has reversed cause and effect.

Example 22 Zoe: My life's a mess. I've never really been happy since all those years ago in school you told Sally that I hated her cat. She believed your stupid joke and made sure I wasn't a cheerleader. I'll never be a cheerleader. It's your fault I'm so miserable now.

Dick: There, there.

Analysis Zoe is tracing the cause too far back. Dick rightly doesn't try to reason with her because he remembers the Principle of Rational Discussion.

Example 23 Money causes counterfeiting.

Analysis This is a general causal claim covering every particular claim like "That there was money in this society caused this person to counterfeit the currency." We certainly have lots of evidence. The problem seems to be that though this is true, it's uninteresting. It's tracing the cause too far back. There being money in a society is part of the normal conditions when we have the effect that someone counterfeited currency.

Example 24 "When more and more people are thrown out of work, unemployment results."   President Calvin Coolidge

   Analysis This isn't cause and effect, it's a definition.

Example 25 (Advertisement by the Iowa Egg Council in the Des Moines Iowa International Airport) "Children who eat breakfast not only do better academically, but they also behave better."

                                 Archives of Pediatric and Adolescent Medicine

   Analysis They're hoping you'll believe that the correlation means there's cause and effect. But you know to look for other possibilities. In particular, there could be a common cause: their parents are richer and/or spend more time with them, which is why they get breakfast and do better academically and behave better.

Example 26 Maria: Fear of getting fired causes me to get to work on time.

   Analysis What is fear? The purported cause here is "Maria is afraid of getting fired," the effect: "Maria gets to work on time."

   Is it possible for Maria to be afraid of getting fired and still not get to work on time? Certainly, but not, perhaps, under normal conditions: Maria sets her alarm; the electricity doesn't go off; the weather isn't bad; Maria doesn't oversleep; . . . .

   But doesn't the causal claim mean it's because she's afraid that Maria makes sure these claims will be true, or that she'll get to work even if one or more is false? She doesn't let herself oversleep due to her fear. In that case how can we judge whether what Maria said is true? It's easy to think of cases where the cause is true and effect false. So we have to add normal conditions. But that Maria gets to work regardless of conditions that aren't normal is what makes her consider her fear to be the cause.

   Subjective causes are often a matter of feeling, some sense that we control what we do. They are often too vague for us to classify as true or false.

Example 27 [Advertisement] "Studies have shown that three cups of Cheerios® a day with a low-fat diet can help lower cholesterol."

Analysis A low-fat diet by itself will help reduce cholesterol, so it's not clear that the purported cause makes a difference. And three cups of Cheerios is a lot of Cheerios.

Example 28 Lee: My neighbor said it's been the worst season ever for allergies this spring, but I told her I hadn't had any bad days. Then today I started sneezing. Darn it—if only she hadn't told me.

Analysis This may be cause and effect, but the evidence shouldn't convince. It's just *post hoc ergo propter hoc*.

Example 29 Dick: Hold the steering wheel.
     Zoe: What are you doing? Stop! Are you crazy?
     Dick: I'm just taking my sweater off.
     Zoe: I can't believe you did that. It's *so* dangerous.
     Dick: Don't be silly. I've done it a thousand times before.
     Crash . . . Later
     Dick: You had to turn the steering wheel!? That made us crash.

Analysis The purported cause is that Zoe turned the steering wheel; the effect is that the car crashed. The necessary criteria are satisfied. But Zoe's turning the steering wheel is a *foreseeable consequence* of Dick making her take the wheel, which is the real cause. The normal conditions are not just what has to be true before the cause, but also what will normally follow the cause.

Example 30 The Treaty of Versailles caused World War II.

Analysis The purported cause is "The Treaty of Versailles was agreed to and enforced." The purported effect is "World War II occurred." To analyze a conjecture like this, an historian will write a book. The normal conditions have to be spelled out. She has to show that it was a foreseeable consequence of the enforcement of the Treaty of Versailles that Germany would re-arm. But was it foreseeable that Chamberlain would back down over Czechoslovakia? More plausible is that the signing of the Treaty of Versailles is *a* cause, not *the* cause of World War II. When several claims together are taken *jointly* as the cause, we say that each describes *a cause* or is a *causal factor*.

Example 31 Tom: The only time I've had a really bad backache is right after I went bicycling early in the morning when it was so cold last week. Bicycling never bothered me before. So it must be the cold weather that caused my back to hurt after cycling.

Analysis   The purported cause is "It was cold when Tom went cycling," the effect is "Tom got a backache." The criteria seem to be satisfied. But Tom may have overlooked another cause. He also had an upset stomach, so maybe it was the flu. Or maybe it was tension, since he'd had a fight with Suzy the night before. He'll have to try cycling in the cold again to find out. Even then he may be looking too hard for *the* cause, when there may be several causes jointly. Another possibility: Tom will never know for sure.

Example 32   Dick: Wasn't it awful what happened to old Mr. Grz?
Zoe: You mean those tree trimmers who dropped a huge branch on
        him and killed him?
Dick: You only got half the story. He'd had a heart attack in his car
        and pulled over to the side. He was lying on the pavement when
        the branch hit him and would have died anyway.
    Analysis   What's the cause of death? Mr. Grz would have died anyway. So the tree branch falling on him wouldn't have made a difference. But the tree branch falling on him isn't a foreseeable consequence, part of the normal conditions of his stumbling out of his car with a heart attack. It's an ***intervening cause***.

Example 33   Poltergeists are making the pictures fall down from their hooks.
    Analysis   To accept this, we have to believe that poltergeists exist. That's dubious. Worse, it's probably not *testable*: How could you determine if there are poltergeists? Dubious claims that aren't testable are the worst candidates for causes.

Example 34   Running over nails causes your tires to go flat.
    Analysis   This sounds right, but it's false. Lots of times we run over nails and our tires don't go flat. What's correct is: "Running over nails *can cause* your tires to go flat." That is, if the conditions are right, running over a nail will cause your tire to go flat. In the next chapter we'll look at the difference between "causes" and "can cause."

Example 35   "Since he was adopted by staff members as a kitten, Oscar the Cat has had an uncanny ability to predict when residents are about to die. Thus far, he has presided over the deaths of more than 25 residents on the third floor of Steere House Nursing and Rehabilitation Center in Providence, Rhode Island. His mere presence at the bedside is viewed by physicians and nursing home staff as an almost absolute indicator of impending death, allowing staff members to adequately notify families. Oscar has also provided companionship to those who would otherwise have died alone. For his work, he is highly regarded by the physicians and staff at Steere House and by the families of the residents whom he serves."

David M. Dosa, M.D., *New England Journal of Medicine*, 7/26/2007

Analysis This is very sweet and mysterious. How does Oscar the Cat know the person is going to die? But reversing cause and effect is just as plausible given the evidence: Oscar the Cat visiting the person causes the person to die. Mysteries merit further investigation, not slack-jawed credibility.

18 Cause in Populations

Cause-in-population studies

When we say smoking causes lung cancer, what do we mean? If you smoke a cigarette, you'll get cancer? If you smoke a lot of cigarettes this week, you'll get cancer? If you smoke 20 cigarettes a day for 40 years, you'll get cancer? It can't be any of these, since there are lots of people who smoke who did all that yet didn't get lung cancer, and the effect has to (almost) invariably follow the cause.

Cause in a population is usually explained as meaning that given the cause, there's a higher probability that the effect will be true than if the cause had not occurred. In this example, people who smoke have a much higher probability of getting lung cancer. But really we're talking about cause and effect just as we did before. Smoking lots of cigarettes over a long period of time will cause (inevitably) lung cancer. The problem is that we can't state, we have no idea how to state, nor is it likely that we'll ever be able to state the normal conditions for smoking to cause cancer. Among other factors, there are diet, where one lives, exposure to pollution and other carcinogens, and one's genetic inheritance. But *if we knew exactly* we'd say: "Under the conditions _____ , smoking ___ (number of) cigarettes every day for ___ years will result in lung cancer."

Since we can't specify the normal conditions, the best we can do is point to the evidence that convinces us that smoking is a cause of lung cancer and get an argument with a statistical conclusion: "People who continue to smoke two packs of cigarettes per day for 10 years are __% more likely (with margin of error __%) to get lung cancer."

How do we establish cause in a population?

Controlled experiment: cause-to-effect

This is our best evidence. We choose 10,000 people at random and ask 5,000 of them never to smoke and 5,000 of them to smoke a pack of cigarettes every day. We have two samples, one composed of those who are administered the cause, and one of those who are not, the latter called the **control group**. We come back 20 years later to check how many in each group got lung cancer. If a lot more of the smokers got lung cancer, and the groups were representative of the population as a whole, and we can see no other *common thread* among those who got lung cancer, we'd be justified in saying that smoking causes lung cancer.

Of course such an experiment would be unethical, so we use rats or cats instead and then argue by analogy. Whether that's more ethical is another issue.

Uncontrolled experiment: cause-to-effect Here we take two randomly chosen samples of the general population for which we have factored out other known possible causes of lung cancer, such as working in coal mines. One of the groups is composed of people who say they never smoke. The other group is composed of people who say they smoke. We follow the groups and 20 years later check whether those who smoked got lung cancer more often. Since we think we've accounted for other common threads, smoking is the remaining common thread that may account for why the second group got cancer more often.

This is a cause-to-effect experiment, since we start with the suspected cause and later see if the effect followed. But it is uncontrolled: some people may stop smoking, some may begin, people have quite variable diets—there may be a lot we'll have to factor out in trying to assess whether it's smoking that causes the extra cases of lung cancer.

Uncontrolled experiment: effect-to-cause Here we look at as many people as possible who have lung cancer to see if there is some common thread that occurs in (almost all) their lives. We factor out those who worked in coal mines, we factor out those who lived in high pollution areas, those who have cats, ... If it turns out that a much higher proportion of the remaining people smoked than in the general population, we have good evidence that smoking was the cause (the evaluation of this requires a knowledge of statistics). This is uncontrolled because how they got to the effect was unplanned, not within our control. And it is an effect-to-cause experiment because we start with the effect in the population and try to account for how it got there.

Examples

Example 1 Barbara smoked two packs of cigarettes a day for 30 years. Barbara now has lung cancer. Barbara's smoking caused her lung cancer.

Analysis   Is it possible for Barbara to have smoked two packs of cigarettes each day for 30 years and not get lung cancer? We can't state the normal conditions. So we invoke the statistical relation between smoking and lung cancer to say it is unlikely for the cause to be true and effect false.

Does the cause make a difference? Could Barbara have gotten lung cancer even if she had not smoked? Suppose we know that she wasn't a coal miner, didn't work in a textile factory, didn't live in a city with a very polluted atmosphere, and didn't live with a heavy smoker, all conditions that are known to be associated with a higher probability of getting lung cancer. Then it is possible for Barbara to have gotten lung cancer anyway, since some people who have no other risks do get lung cancer. But it is very unlikely, since very few of those people do.

We have no reason to believe that there is a common cause. Maybe people with a certain biological make-up feel compelled to smoke, and that biological make-up also contributes to their getting lung cancer independently of their smoking. But we've no evidence for that, and before cigarette smoking became popular lung cancer was rare.

So assuming a few normal conditions, "Barbara's smoking caused her lung cancer" is as plausible as the strength of the statistical link between smoking and lung cancer and the strength of the link between not smoking and not getting lung cancer. We must be careful, though, that we do not attribute the cause of the lung cancer to smoking just because we haven't thought of any other cause, especially if the statistical link isn't very strong.

Example 2   Zoe: I can't understand Melinda. She's pregnant and she's drinking.

Dick: That's all baloney. I asked my mom, and she said she drank when she was pregnant with me. And I turned out fine.

Zoe: But think how much better you'd have been if she hadn't.

Analysis   Zoe doesn't say but alludes to the cause-in-population claim that drinking during pregnancy causes birth defects or poor development of the child. That has been demonstrated: many cause-in-population studies have been done that show there is a higher incidence of birth defects and developmental problems in children born to women who drink during pregnancy than to women who do not drink, and those defects and problems do not appear to arise from any other common factor.

Dick, however, makes a mistake. He confuses a cause-in-population claim with a general causal claim. He's right that his mother's experience would disprove the general causal claim, but it has no force against the cause-in-population claim.

Zoe's confusion is that she thinks there is a perfect correlation between drinking and physical or mental problems in the child, so that if Dick's mother had not drunk he would have been better, even if Zoe can't point to the particular way in which Dick would have been better. But the correlation isn't perfect; it's only a statistical link.

Example 3   Lack of education causes poverty. Widespread poverty causes crime. So lack of education causes crime.

Analysis   We often hear words like these, and some politicians base policy on them. But they're too vague. How much education constitutes "lack of education"? How poor do you have to be? How many poor people constitute "widespread poverty"? Researchers make these sentences more precise and analyze them as cause-in-population claims, since we know they couldn't be true general causal claims: there are people with little education who've become rich; and lots of poor people are law-abiding citizens. Indeed, during the worst years of the Depression in the 1930s, when there was more widespread poverty than at any time since in the U.S., there was less crime than any time in the last 20 years. This suggests it would be hard to find a precise version of the second sentence that is a true cause-in-population claim.

Example 4   "The number of teenagers giving birth declined 2 percent in the United States in 2008, reversing two years of increases, as older teens may have delayed starting a family because of the recession."

Albuquerque Journal, 4/7/2010

Analysis   The author conjectures a cause-in-population claim on nothing more than *post hoc* evidence.

Example 5   A little booze does a woman's mind good
"Women who imbibe a little wine, beer or spirits every day are less likely than teetotalers to see their memories and other thinking powers fade as they age, according to the largest study to assess alcohol's impact on the brain. The study of more than 12,000 elderly women found that those who consumed light to moderate amounts of alcohol daily had about a 20 percent lower risk of experiencing problems with their mental abilities later in life.

'Low levels of alcohol appear to have cognitive benefits,' said Francine Grodstein of the Brigham and Women's Hospital in Boston, senior author of the study, which is being published in today's New England Journal of Medicine. 'Women who consistently were drinking about one-half to one drink per day had both less cognitive impairment as well as less decline in their cognitive function compared to women who didn't drink at all.'

While the study involved only women, the findings probably hold true for men, although previous research indicates that men seem to benefit from drinking slightly more—one to two drinks per day, researchers said.

The findings provide the latest evidence that indulging in alcohol, long vilified as part of an insalubrious lifestyle, can actually help people live longer, healthier lives. While heavy drinking clearly causes serious problems for many people, recent research has found that drinking in moderation protects the heart."   Washington Post, 1/15/2005

Analysis Correlation does not establish cause and effect. It could be the reverse here: elderly women who are mentally alert prefer to have something to drink to slow them down to sleep better. Or there could be a common cause. It's not even clear from this article whether this was a cause-to-effect or effect-to-cause study. More studies are needed, at least from the little we learn in this write-up.

Example 6 "The US Bureau of Labor Statistics data from 2001 show the following:

Education and Lifetime Income

Highest Education Level Achieved	*Lifetime Income* (40 years)
Bachelor's Degree	$1,667,700
Associate Degree	$1,269,850
High School Graduate	$994,080
Not High School Graduate	$630,000

Higher levels of education payoff in lifetime income in a big way.

It is interesting to note that this relationship between education and earnings potential has been known since the 1970's, and has been consistently demonstrated by government surveys. In fact the difference in income level with education has grown significantly over the years. The Bureau of the Census has suggested that the gap in earnings between those with higher education and those with lower education will continue to grow in the future. The US Bureau of Labor Statistics has also shown that the unemployment rate steadily drops with higher levels of education. Unemployment for non-high school graduates was 6.5% in 2000, 3.5% for high school graduates, and 2.3% for those with an associate degree.

Education makes a difference! "

Education Online, 2010, www.education-online-search/articles/special_topics/education_and_income/.com

Analysis There's a clear correlation between income and level of education. The website claims that this means getting more education is the cause of earning more ("payoff," "education makes a difference"). But people who finish more schooling are brighter, are either wealthier or can figure out how to get money for their education, are willing to work hard and to persevere. People like that are likely to earn more than other folks whether they get more education or not. Without more evidence, more studies that factor out these possible common causes, this is just *post hoc* reasoning.

Example 7 "In the mid-1970s a team of researchers in Great Britain conducted a rigorously designed large-scale experiment to test the effectiveness of a treatment program that represented 'the sort of care which today might be provided by most specialized alcoholism clinics in the Western world.'

The subjects were one hundred men who had been referred for alcohol problems to a leading British outpatient program, the Alcoholism Family Clinic of Maudsley Hospital in London. The receiving psychiatrist confirmed that each of the subjects met the following criteria: he was properly referred for alcohol problems, was aged 20 to 65 and married, did not have any progressive or painful physical disease or brain damage or psychotic illness, and lived within a reasonable distance of the clinic (to allow for clinic visits and follow-up home visits by social workers). A statistical randomization procedure was used to divide the subjects into two groups comparable in the severity of their drinking and their occupational status.

For subjects in one group (the 'advice group'), the only formal therapeutic activity was one session between the drinker, his wife, and the psychiatrist. The psychiatrist told the couple that the husband was suffering from alcoholism and advised him to abstain from all drink. The psychiatrist also encouraged the couple to attempt to keep their marriage together. There was a free-ranging discussion and advice about the personalities and particularities of the situation, but the couple was told that this one session was the only treatment the clinic would provide. They were told in sympathetic and constructive language that the 'attainment of the stated goals lay in their hands and could not be taken over by others.'

Subjects in the second group (the 'treatment group') were offered a year-long program that began with a counseling session, an introduction to Alcoholics Anonymous, and prescriptions for drugs that would make alcohol unpalatable and drugs that would alleviate withdrawal suffering. Each drinker then met with a psychiatrist to work out a continuing outpatient treatment program, while the social worker made a similar plan with the drinker's wife. The ongoing counseling was focused on practical problems in the areas of alcohol abuse, marital relations, and other social or personal difficulties. Drinkers who did not respond well were offered in-patient admissions, with full access to the hospital's wide range of services.

Twelve months after the experiment began, both groups were assessed. No significant differences were found between the two groups. Furthermore, drinkers in the treatment group who stayed with it for the full period did not fare better than those who dropped out. At the twelve-month point, only eleven of the one hundred drinkers had become abstainers. Another dozen or so still drank but in sufficient moderation to be considered 'acceptable' by both husband and wife. Such rates of improvement are not significantly better than those shown in studies of the spontaneous or natural improvement of chronic drinkers not in treatment."

Herbert Fingarette, *Heavy Drinking:The Myth of Alcoholism as Disease*

Analysis The controlled cause-to-effect study reported on here appears to be very well done.

Reasoning
in the
Sciences

19 The Scientific Method

The best way to determine cause and effect is to experiment.

Example 1 I have a waterfall in my backyard in Cedar City. The
pond has a thick, rubberized, plastic pond liner, and I have a pump and
hose that carry water from the pond along the rock face of a small rise
to where the water spills out and runs down more rocks with concrete
between them. Last summer I noticed that the pond kept getting low
every day and had to be refilled. You don't waste water in the desert,
so I figured I'd better find out what was causing the loss of water.

I thought of all the ways the pond could be leaking: the hose that
carries the water could have a leak, the valve connections could be
leaking, the pond liner could be ripped (the dogs get into the pond to
cool off in the summer), there could be cracks in the concrete, or it
could be evaporation and spray from where the water comes out at the
top of the fountain.

I had to figure out which (if any) of these was the problem. First I
got someone to come in and use a high-pressure spray on the waterfall
to clean it. We took the rocks out and vacuumed out the pond. Then
we patched every possible spot on the pond liner where there might be
a leak.

Then we patched all the concrete on the waterfall part and water-
sealed it. We checked the valve connections and tightened them. They
didn't leak. And the hose wasn't leaking because there weren't any wet
spots along its path.

Then I refilled the pond. It kept losing water at about the same rate.

It wasn't the hose, it wasn't the connections, it wasn't the pond
liner, it wasn't the concrete watercourse. So it had to be the spray and
evaporation.

I reduced the flow of water so there wouldn't be so much spray.
There was a lot less water loss. The rest I figured was probably evaporation, though there might still be small leaks.

In trying to find the cause of the water leak at my waterfall and
pond, I was using a method scientists often use.

The scientific method Conjecture possible causes. By experiment eliminate them by showing they don't make a difference until there is only one. Check that one: Does it make a difference? If the purported cause is eliminated, is there still the effect? Is there a common cause?

I assumed there was a cause, then by a process of elimination on some conjectured causes, I fixed on one: when that occurred, the effect always did, too, and it made a difference, and I knew I could fill in the normal conditions.

But why should I assume there is a cause? Does this mean I'm assuming everything has a cause? No, I'm assuming that there is some way to stop the leak, which in this case amounts to assuming that the leak has a cause. The assumption that a particular effect has a cause is sometimes just an expression of our desire to find a way to manipulate the world.

But then doesn't this method rest on a false dilemma?

A or B or C is the cause of E. It's not A. It's not B.
Therefore, it's C.

No. We also have to check that C satisfies all the conditions for cause and effect, not just that it makes a difference. We must be willing to accept that our experiments will show that none of the conjectured causes satisfies all the conditions. This method cannot find the cause from nothing, but only, if we guess right, isolate it from a range of conjectured causes.

Example 2 Recently Lee found out that he has hepatitis B. None of his friends has hepatitis. He wonders how he could have gotten it.

He reasons: Since he wants to be a nurse, he volunteers to work at a hospital three times per week. Some of the patients there have hepatitis, and he often washes their bedpans and comes in contact with their body fluids, though he's always careful to wear gloves. Or at least he thought he was. A recent study he read said that 25% of health-care workers exposed to hepatitis B get it. So, he figures, he got hepatitis B from working at the hospital.

Analysis How strong is this argument? At best we can say that "Lee contracted hepatitis B from working at the hospital" is a good conjecture. We rule out all other causes we can think of. We can imagine conditions under which he could have gotten hepatitis there, but we can't specify the exact conditions that occurred that would give us the normal conditions. Eliminating all other possible causes (that we can think of) doesn't mean that we can conclude we've found the cause unless we also have:

(*) The only ways Lee could have gotten hepatitis B
     are P, Q, R, S, T, U, or V.

There are very strong arguments that he didn't get it from Q, R, S, T, U, or V. Therefore (reasoning by excluded middle), he got it from P. This reasoning to a cause is just as strong as (*) is plausible.

Example 3 "In my backyard, indeed throughout the neighborhood where I live, the abundance of birds is limited. In other neighborhoods there are many more birds. The most important difference I can think of concerns cats. Many cats are around where I live; elsewhere there are less of them. It is probable that there will be other differences between neighborhoods which differ in bird abundance. However, in view of background information it is reasonable to infer that cats will be a causal factor. Cats eat birds and birds are afraid of cats. An experiment could provide more confirmation. If I would shoot the cats near my place and bird abundance would subsequently increase, I would feel confident that cats do influence the abundance of birds. . . .

   If an experiment of this kind were indeed performed with positive results (for the birds I mean), the evidence would be telling. However, we should realize that the situations compared—before and after the shooting—may differ in other respects. Thus it is possible that, from a bird's point of view, there happens to be a long-lasting improvement of the weather after the shooting.

   In view of this the following experiment would be more decisive. Suppose we identify ten neighborhoods with many cats. We could remove the cats from five randomly chosen neighborhoods, and let the cats be in the remaining ones. If bird abundance would increase in the cat-free areas, not elsewhere, that would be something. It is improbable that the two groups of neighborhoods will systematically differ in another factor that influences birds."

   Wim J. van der Steen, *A Practical Philosophy for the Life Sciences*

These examples test particular causal claims. Scientists, though, are usually interested in general causal claims. To get strong arguments for those, we use cause-in-population studies.

20 Experiments

What counts as evidence in science?

> **Observational claim** An observational claim is one that is established by personal experience or observation in an experiment.
>
> **Evidence** Evidence is usually the observational claims that are used as premises of an argument. Sometimes the term refers to all the premises.

What do we mean by "observation in an experiment"?

A physicist may say that she saw an atom traverse a cloud chamber, when what she actually saw was a line made on a piece of photographic film. A biologist may say he saw the nucleus of a cell, when what he saw was an image projected through a microscope. In both cases these people are not reporting on direct personal experience but on deductions made from that personal experience. However, those claims made by deduction from the perceptions arising from certain types of experiments are, by consensus in that area of science, deemed to be observations.

Within any one area of science there is a high level of agreement on what counts as an observational claim. But from one area of science to another that standard may vary. A physicist beginning work in biology may well question why certain claims are taken as "obvious" deductions from experience, such as the reality of what you see through a microscope. But after the general form of the inference—from such direct claims about personal experience to the observational claims—is made explicit once or twice, he is likely to accept such claims as undisputed evidence. If he doesn't accept such deductions, he is questioning the basis of that science.

When new techniques are introduced into a science or when a new area of science is developing, there is often controversy about what counts as an observational claim. Galileo's report of moons around Jupiter was received with considerable skepticism because telescopes were not assumed to be accurate, and indeed at that time they distorted a lot. In ethology, the study of animal behavior in natural settings, there is no agreement yet on what counts as an observational claim, and you can find different journal articles using different standards. Consider:

Some would describe this as the first chimpanzee getting angry and chasing the second one away, and then the second returning to pacify and re-establish bonds with the first. That's what they saw. But others say that such a description is loaded with assumptions that have not been established, such as that chimpanzees have emotions sufficiently similar to humans to label as anger, and that chimpanzees intend to accomplish certain ends, rather than just operating instinctually.

One constraint we impose on reports of observations is that they should be replicable. We believe that nature is uniform. What can happen once can happen again, *if* the conditions are the same. Scientists typically won't accept reports on observations that they are unable to reproduce.

Duplicable and replicable experiments An experiment is *duplicable* if it is described clearly enough that others can follow the method to obtain observations. It is *replicable* if, when it is duplicated, the observations of the new experiment are in close agreement with the observations of the original experiment.

The difficulty is to specify exactly what conditions are required and what counts as close enough agreement. It's fairly easy in chemistry and physics; less so in biology; much more difficult in psychology or ethology. It's virtually impossible in history and economics, which means history and economics are not sciences, except to the extent that we can describe very general conditions that may recur.

Some examples will illustrate these ideas.

Examples

Example 1 A recipe from a famous coffeehouse
"*Vegetarian Chile*
2 cans each (include liquid): Pinto beans, Chili beans, Great
      Northern beans, Red beans, Kidney beans
    1 # 10 can diced Tomatoes
    Garlic, 6-8 cloves chopped
    Bell Pepper, 1 chopped
    Jalapeño Peppers, 3 chopped
    Chili Powder 2 soup spoons
    Onions, 2 chopped or in food processor
    Paprika, 1 soup spoon

Put in soup tureen and heat to boil for 1 hour. Take care the beans
don't stick to the bottom."

    Analysis Any expert in the subject (any person who has worked
in a commercial kitchen) will know what a #10 can of tomatoes is.
Though "chopped" and "soup spoon" may be unclear, anyone who saw
the chile being made would be able to duplicate the preparation.

Example 2 Feeding behavior of primates
"General Methodology
Data were collected simultaneously on both the activity of the animals
and the forest strata at which this activity took place. Counts were
made at five-minute intervals of the numbers of individuals engaged in
each of the six activities and the level of the forest in which the activity
was performed. The following activities were recorded: feeding—the
animal actually in the process of ingesting or picking a food item;
grooming—mutual and self-grooming were distinguished for certain
analyses; resting—no body displacement, or feeding, or grooming,
sunning, etc.; moving—movement of an individual, including
individual foraging; travel—movement of the group; and other—
e.g., sunning, play, fighting. These data were collected only after the
animals under observation were reasonably habituated to the observer.
Each observation of an animal constituted an individual activity record
(IAR) collected in a given five-minute time sample. Because of the
focus of the study and the difficulty in keeping continuous contact with
an individual animal, no attempt was made to follow individual animals
nor to collect statistical data on specific age or sex classes. Statistical

analyses of the data were complicated by the fact that some of the activity records were not independent of each other. The methods used for the statistical analyses are reported in Sussman *et al.*

To determine levels of the forest, I used Richards' (1957) categories of forest stratification as a model and assigned numbers of one to five to the forest layers. Level 1 is the ground layer of the forest; it includes the herb and grass vegetation. Level 2 is the shrub layer, from one to three metres above the ground. This layer is usually found in patches throughout the continuous canopy forest, but is much more dense and is the dominant layer in the brush and scrub regions. Level 3 of the forest consists of small trees, the lower branches of larger trees, and saplings of the larger species of trees. This layer is about three to seven metres high. Level 4 is the continuous or closed canopy layer. It is about five to 15 metres high. The dominant tree of the closed canopy, at all three forests, is the kily (*Tamarindus indica*). Level 5 of the forest is the emergent layer and consists of the crowns of those trees which rise above the closed canopy. It is usually over 15 metres high.

All three forests in which I made intensive studies were primary forests and the tree layers were quite distinct. In most cases, the particular level in which an animal was observed could be distinguished easily. If I could not determine the forest level unambiguously, I did not record it.

Observations recorded in this manner may be biased because animals that are active in certain levels of the forest may be more difficult to see than those active at other levels. I attempted to minimize this problem by following a relatively small number of animals (usually from five to ten) throughout a period of continuous observation, keeping track of all the animals. For *Lemur fulvus* this usually included the whole group, which was small and, for the most part, moved together. It was more difficult to do this when observing *Lemur catta*, for which it was often necessary to follow and observe subgroups of the larger group. The larger group would disperse, especially during foraging and feeding, and during afternoon rest periods.

Day ranges were mapped by following a group from one night resting site in the morning to the time it settled in another night resting site in the evening. The location of the group was plotted throughout the day on a prepared map of the forest and the amount of time the group spent in each location was recorded. Home ranges include the sum of all the day ranges. The data on home ranges are limited, however, and probably do not represent total home ranges of the groups, since the study in each area was limited to a few months."

R.W. Sussman, "Feeding behaviour of Lemur Catta and Lemur Fulvus" in *Primate Ecology*, ed. T. H. Clutton-Brock

Analysis It is difficult to be more precise than this in ethology. The description of the methodology is clear enough to count as duplicable, perhaps even by someone who isn't an expert in the subject. Whether the observations are replicable will depend on how closely we expect them to agree with the ones in this paper.

Note that the author has not stated what time of year the observations were made, nor the percentage of males versus females in the groups he studied. These are not part of the conditions that need to be duplicated; implicitly, the author is saying they don't matter. If it turns out in trying to duplicate this experiment that different observations are obtained at different times of the year, then the time of year would have to be added as part of the conditions that are important and which have to be duplicated.

Example 3 Cyclic Variations in Grass Growth
"Grass exhibits a cyclical growth pattern surprisingly different from any other known plant. In this study, average grass blade heights have been measured, on a daily basis, over a 10 week period. Measurements were taken, utilizing vernier calipers, of the height of one hundred individual grass blades randomly chosen in a 10 foot square area positioned in front of an apartment complex in the Lexington, Kentucky area. (Measurements were also repeated with a different set of calipers to ensure reproducibility on a different apparatus.) The average of these measurements was computed and experimental error was taken as the standard deviation of the mean divided by the square root of the number of grass blades in the average. The procedure was repeated on a daily basis for a period of 10 weeks.

Figure 1: Experimental measurements of average grass height are plotted versus time. Solid line represents experimental data. Short dashed line indicates a 'constant grass height' calculation and is normalized to the experimental data to produce the best fit.

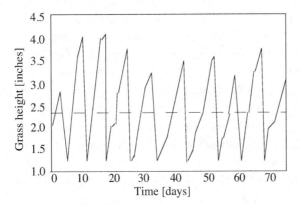

Results and Discussion The average grass heights, measured in this work, are plotted as a function of time in Figure 1. As one can readily see, there exists a periodic variation in average grass height with an approximate cycle of 7 to 10 days. Another intriguing observation is that there exists a minimum grass height, or 'grass baseline,' of about 1.3 inches.

Since the cyclic period of the grass is 7 to 10 days, one may conclude that grass height varies on a 'week-about' basis. The physical mechanism responsible for this cyclic grass height phenomenon is not clearly understood at this time."

V. D. Irby, M. S. Irby, Dept. of Physics and Astronomy, University of
Kentucky, *Annals of Improbable Research*, vol. 1, no. 4, 1995

Analysis The authors take great care that their experiment can be duplicated, and almost certainly it is replicable. But you should realize by now that this doesn't make it a good experiment.

Example 4   The refraction of light rays

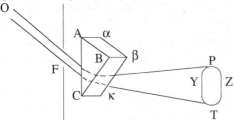

"In the wall or window of a room let F be some hole through which solar rays OF are transmitted, while other holes elsewhere have been carefully sealed so that no light enters from any other place. The darkening of the room, however, is not necessary; it only enables the experiment to turn out somewhat more clearly. Then place at that hole a triangular glass prism AαBβCκ that refracts the rays *OF* transmitted through it toward *PYTZ*."

Isaac Newton, *Optica*, Part 1, Lecture 1, 1670, translated from the Latin
in *The Optical Papers of Issac Newton,* ed. Alan E. Shapiro

Analysis This is very clear because of the diagram. It can be and often was duplicated, and the observations were replicable.

Example 5 Testing for anomalous cognition (ESP)
"The vast majority of anomalous cognition experiments at SRI [Stanford Research Institute] and SAIC [Science Applications International Corporation] used a technique known as remote viewing. In these experiments, a viewer attempts to draw or describe (or both) a target location, photograph, object, or short video segment. All known channels for receiving the information are blocked. Sometimes the viewer is assisted by a monitor who asks the viewer questions; of course, in such cases the monitor is blind to the answer as well. Sometimes a sender is looking at the target during the session, but sometimes there is no sender. In most cases the viewer eventually receives a feedback in which he or she learns the correct answer, thus making it difficult to rule out precognition [knowing the future] as the explanation for positive results, whether or not there was a sender.

Most anomalous cognition experiments at SRI and SAIC were of the free-response type, in which viewers were asked simply to describe the target. ...

The SAIC remote-viewing experiments and all but the early ones at SRI used a statistical evaluation method known as rank-order judging. After the completion of a remote viewing, a judge who is blind to the true target (called a blind judge) is shown the response and five potential targets, one of which is the correct answer and the other four of which are 'decoys.' Before the experiment is conducted, each of those five choices must have had an equal chance of being selected as the actual target. The judge is asked to assign a rank to each of the possible targets, where a rank of 1 means it matches the response most closely, and a rank of 5 means it matches the least.

The rank of the correct target is the numerical score for that remote viewing. By chance alone the actual target would receive each of the five ranks with equal likelihood, since, despite what the response said, the target matching it best would have the same chance of selection as the one matching it second best and so on. The average rank by chance would be 3. Evidence for anomalous cognition occurs when the average rank over a series of trials is significantly lower than 3. (Notice that a rank of 1 is the best possible score for each viewing.)

This scoring method is conservative in the sense that it gives no extra credit for an excellent match. A response that describes the target almost perfectly will achieve the same rank of 1 as a response that contains only enough information to pick the target as the best choice out of the five possible choices."

Jessica Utts, "An assessment of the evidence for psychic functioning"
The Journal of Parapsychology, vol. 59, no. 4, p. 289, 1995

Analysis What does "All known channels for receiving information are blocked" mean? We need to know the exact layout of the room where the experiment was done. "In most cases the viewer eventually receives feedback"—how often, under what circumstances, exactly when? We need to know how close the "decoys" were to the true target. Who are the judges? This is crucial because different judges from different backgrounds may classify differently.

The experiment is not duplicable. Even if you watched the experiment being done you couldn't duplicate it, for it's not clear what the author considers important and what she considers irrelevant in the setup.

Even if it were possible to duplicate the experiment and get the same results, it's not clear that by chance alone the actual target would not receive each of the five ranks with equal likelihood. Perhaps this experiment would show the opposite.

Example 6   The growth of living nerve cells in vitro
"The immediate object of the following experiments was to obtain a method by which the end of a growing nerve could be brought under direct observation while alive, in order that a correct conception might be had regarding what takes place as the fibre extends during embryonic development from the nerve center out to the periphery.

The method employed was to isolate pieces of embryonic tissue known to give rise to nerve fibres, as for example, the whole or fragments of the medullary tube or ectoderm from the branchial region, and to observe their further development. The pieces were taken from frog embryos about three mm. long, at which stage, i.e. shortly after the closure of the medullary folds, there is no visible differentiation of the nerve elements. After carefully dissecting it out the piece of tissue is removed by a fine pipette to a cover slip upon which is a drop of lymph freshly drawn from one of the lymph sacs of an adult frog. The lymph clots very quickly, holding the tissue in a fixed position. The cover slip is then inverted over a hollow slide and the rim sealed with paraffin. When reasonable aseptic precautions are taken, tissues will live under these conditions for a week and in some cases specimens have been kept alive for nearly four weeks. Such specimens may be readily observed from day to day under highly magnifying powers."

> Ross Harrison, *Proceedings of the Society for Experimental and Medicine Biology*, vol. 4, 1907 (as quoted in *The Origins and Growth of Biology*, ed. Arthur Rook, pp. 159–160)

Analysis This is the first method ever recorded for maintaining living cells outside the body. It is very much like the recipe from the *Dog & Duck*. Even for an expert it would have been difficult to duplicate it from just reading this.

The morals of these examples

- It's very hard to describe an experiment clearly enough to duplicate it.

- What is described in an experiment is what needs to be duplicated. What is not described is deemed irrelevant to obtaining similar observations.

- What counts as duplicable is going to be relative to the particular scientific discipline. Expert knowledge in the area may make some descriptions very clear.

- What counts as close enough agreement for observations to be deemed replicable is going to depend on the particular scientific discipline.

- New experimental designs are often sketchily described, but they are accepted anyway because people go to the lab, see how it is done, then go back to their labs and do the experiment, and then pass that on to other people.

21 Explanations

Inferential explanations

Why does the sun rise in the east? How does electricity work? How come Spot gets a bath every week? Why didn't you give me an A on the last exam? We give explanations as answers to lots of different kinds of questions. Our answers can be as varied as the questions. We can tell a story, a myth about how the world was created. We can write a scientific treatise on how the muscles of the esophagus work. We can give instructions for how to play a guitar. We can draw a map.

Here we'll focus on verbal explanations. To begin, let's consider ones that answer the question "Why is this true?"

Explanations An *explanation* is a collection of claims that can be understood as *E* because of *A*, *B*, *C*, ... We often call *A*, *B*, *C*, . . . the *explanation* and *E the claim being explained.*

An *inferential explanation* is one that is meant to answer the question "Why is *E* true?"

An inferential explanation is meant to show *why* a claim is true. That's different from an argument to show *that* a claim is true. With an explanation we should already have good reason to believe, say, "The sky is blue," and the explanation should provide us with other claims from which it follows.

Example 1 Zoe: Why is Spot limping?
　　　　Dick: Here, I see. It's because he's got a tiny thorn in his paw.
　　　　Analysis This is an inferential explanation: "Spot has a thorn in his paw" is meant to explain "Spot is limping."

Necessary conditions for an inferential explanation to be good

The claim that's meant to be explained is very plausible
We can't explain what's dubious.

Example 2　Dick: Why is it that most people who call psychic hot lines are women?

Zoe: Wait a minute, what makes you think more women than men call psychic hot lines?

Analysis Dick has posed a loaded question, and Zoe has responded appropriately, asking for an argument to show that "More women than men call psychic hot lines" is true.

The explanation answers the right question

Questions are often ambiguous, and a good explanation to one reading of a question can often be a bad explanation to another. If a question is ambiguous, that's a fault of the person asking the question—we can't be expected to guess correctly what's meant. An explanation is bad because it answers the wrong question *if* it's very clear what question the person is asking.

Analysis Flo gave a good explanation—to the wrong question.

The claims doing the explaining are plausible

In an inferential explanation the claims doing the explaining are supposed to make clear why the claim we are explaining is true. They can't do that if they aren't plausible.

Example 4 The sky is blue because there are blue globules very high in the sky.

Analysis This is a bad explanation because "There are blue globules very high in the sky" is not plausible.

The explanation is valid or strong

The truth of the claim that's being explained is supposed to follow from the claims doing the explaining. So the relation between those claims should be valid or strong, like the relation between the premises and conclusion of a good argument.

Example 5 Dogs lick their owners because dogs aren't cats.

Analysis This is a bad explanation. The relation of "Dogs aren't cats" to "Dogs lick their owners" is neither valid nor strong, and there's no obvious way to repair it.

As with arguments, *we allow that an explanation might need repair*. An explanation *E because of A, B, C, . . .* might require further claims to supplement *A, B, C, ...* But a good inferential explanation will have at least one claim among those that do the explaining that is less plausible than what's being explained. Otherwise it wouldn't explain, it would be a way to convince.

Example 6

Analysis Zoe offers a good explanation of why Dick has a headache:

Anyone who drinks that much is going to have a headache.

Therefore (explains why), Dick has a headache.

Judged as an argument, however, this is bad, for it begs the question: it's a lot more obvious to Dick that he has a headache than that anyone who drinks that much is going to have a headache.

The explanation is not circular

We can't explain why a claim is true by just restating the claim in other words.

Example 7 Zoe: Why can't you write today, Dick?

Dick: Because I've got writer's block.

Analysis This is a bad explanation: "I've got writer's block" just means you can't write.

> **Necessary conditions for an inferential explanation to be good**
> For the inferential explanation *E because of A, B, C, . . .*
> to be good, the following must hold:
>
> • *E* is very plausible.
>
> • *A, B, C, . . .* answer the right question.
>
> • Each of *A, B, C, . . .* is plausible, but at least one of them is not more plausible than *E*.
>
> • The inference *A, B, C, . . . therefore E* is valid or strong, possibly with respect to some other plausible claims.
>
> • The explanation is not *E because of D* where *D* is *E* itself or a simple rewriting of *E*.

Often we say that an explanation is *right* or *correct* rather than "good," and *wrong* rather than "bad." Note that these are necessary conditions for an inferential explanation to be good. There's no agreement on sufficient conditions. That requires judgment.

Causal explanations

When an inferential explanation is given in terms of cause and effect, *if it's good causal reasoning and it answers the right question, then the explanation is good; otherwise it's bad.*

Example 8 Suzy: Why did Dick wake up?
   Zoe: Because Spot was barking.
   Analysis This is a good causal explanation (see p. 112).

Example 9 Dick recovered from his cold in one week because he took vitamin C.
   Analysis This is a causal explanation, but not a good one: the purported cause does not clearly make a difference.

   "Proper treatment will cure a cold in seven days, but left to itself a cold will hang on for a week."—Henry G. Felsen

Example 10 Zoe: I wish I could help Wanda. What's the reason for her weight problem?
   Dick: Gravity.
   Analysis This is a causal explanation, but not a good one. The existence of gravity is a normal condition for Wanda weighing a lot, not the or a cause.

Example 11   Zoe: You say that this argument is bad. But why?
> Dr. E: It's bad because it's weak, for example, Sheila could have
> been a rabbit or a herring.

Analysis Dr. E knows what he's talking about, and this is a good inferential explanation. But it's not a causal one. Explanations in terms of rules or criteria aren't causal.

Examples

Example 12   Customer: Why did you call your coffeehouse *The Dog & Duck*?
> Owner: Why not?

Analysis Shifting the burden of proof is just as bad for explanations as for arguments.

Example 13   Zoe: Why was Tom so unpleasant to us today?
> Dick: Oh, don't mind him. He was just out of sorts.

Analysis This looks circular, but it's worse. Being out of sorts indeed makes people rude—if we had a clear idea of what "being out of sorts" means besides acting rudely. The purported explanation is too vague.

Example 14   Suzy: Why did Dick just get up and leave the room like that in the middle of what Tom was saying?
> Zoe: Because he wanted to.

Analysis This is a bad explanation. Wanting to leave the room when Tom is talking is something unusual and requires further explanation. An explanation is *inadequate* if it leads to a further "Why?" Even if the claims doing the explaining are obviously true, they may not be enough.

Example 15   Dick: Why did that turtle cross the road?
> Lee: Because its leg muscles carried it.

Analysis This is bad because it's the wrong kind of answer. Normally we'd understand Dick to be asking for a behavioral explanation: claims about the motives or beliefs or feelings of a person or creature. What's been given is a premise about the physical makeup of the turtle, a physical explanation. It answers the wrong question.

    On the other hand, "To get to the other side" would be an inadequate explanation. Since we don't know what the turtle's motives are, or indeed if it has any, it's unlikely we can give a good answer to Dick.

Example 16 Psychiatrist: Where is Dr. E? It's time for his
                        appointment.
Secretary: Don't you remember? He said he wouldn't be coming
                        anymore.
Psychiatrist: He is resisting the understanding that I am bringing him.
Secretary: But he says it's because he can't afford your fee.
Psychiatrist: Ah! There, there's the proof that his unconscious is
                        resisting, because I know he could borrow money
                        from his rich uncle.

Analysis Psychiatrists often make their explanations immune to testing. If anything counts as resistance, if everything can be explained in terms of unconscious motives, there's no way to test. We might as well say that a patient won't come because gremlins are inhabiting his psyche, though that might be less effective in getting a patient to continue treatment and pay the bills. *If a claim explains everything, it explains nothing. This is a bad causal explanation. Untestable claims are the worst candidates for a good explanation.*

Example 17 Harry: I can't believe my uncle Ralph gave up his partnership in his law firm. He was making really big bucks. He sold his home and bought a cabin out near Big Tree Meadow, really primitive, and he says he's meditating. He was always such a responsible guy.
     Zoe: How old is he?
     Harry: 45.
     Zoe: He's just going through midlife crisis.

Analysis This is an inferential explanation, but we need more evidence for the claim that he's going through a midlife crisis before we accept it as good—unless you define "midlife crisis" as what Harry's uncle is doing, in which case the explanation is circular.

Example 18 Zoe: My mom is always hot and irritable now.
     Dick: Why's that?
     Zoe: She's going through menopause.

Analysis This is a good causal explanation because there's good cause-in-population evidence that most women who go through menopause have those symptoms.

Example 19 Dick: Why won't Spot eat his dog food today?
     Zoe: He hasn't been hungry all day.
     Dick: I don't think so—he just ate the doggie treat I gave him.

Analysis Dick has shown that Zoe's explanation is not good by showing that the claim doing the explaining is false.

Example 20 The gas has temperature 83° C because it has pressure 7 kg/cm² and volume 807 cm³.
  Analysis This is a good explanation if the claims are true. But it's not causal, because the pressure, volume, and temperature all occur at the same time. Explanations that invoke a law that gives a correlation, such as this one that assumes Boyle's law, are inferential but not causal.

Example 21 Zoe: Why did the lights just go out?
  Dick: The transformer down the street must have blown again. I'll have to call the electric company.
  Zoe: Don't bother—I can see the lights are still on where Flo lives, and the street lights are working.
  Analysis Zoe has used reducing to the absurd to refute Dick's explanation. If Dick's explanation were right, then from the same claim we could conclude, "The street lights and all the lights on the block are out." But that claim is false. So his claim must not be true. So it's a bad explanation.

Example 22 Dr. E: I won't accept your homework late.
  Maria: But I had a meeting I had to attend at work.
  Dr. E: So? I don't count problems with employment as an excuse for handing in late work.
  Analysis Maria has explained why she did not hand in the homework on time. She thinks she's also given an excuse, but Dr. E disabuses her of that. *An explanation is not an excuse.*

Example 23 Dr. Smyrn: Now to check your heart. Do you feel that?
  Dr. E: Ow! Sure. But what has pricking my finger got to do with heart disease?
  Dr. Smyrn: The heart is on the left side of the body, and in heart attacks the victim will normally get a pain in his left arm.
  Dr. E: Gee, that hurts, too.
  Dr. Smyrn: I believe that we can predict heart disease by testing the little finger of the left hand for pain sensitivity. So I compare reactions to pricking the little finger of the left hand to pricking other fingers.
  Dr. E: Really? Ouch. That sounds wacky.

Dr. Smyrn: I have been sending patients with unusual sensitivity in that finger to a cardiologist to follow up.

Dr. E: And? Ow!

Dr. Smyrn: So far he has found indications of heart disease in only 3 of the 32 people I have sent. But that is because my test is more sensitive than his. I can spot that there is incipient heart disease before any other known test.

Analysis This is not an explanation. It's a cause-in-population claim. It could be used to make a causal claim, but no claim is being explained here. Dr. Smyrn's last comments suggest that he does not satisfy the Principle of Rational Discussion, for it looks as if he won't accept any evidence that he's wrong.

Arguments and explanations

Example 24 Dick, Zoe, and Spot are out for a walk in the countryside. Spot runs off and returns after five minutes. Dick notices that Spot has blood around his muzzle. And they both really notice that Spot stinks like a skunk. Dick turns to Zoe and says, "Spot must have killed a skunk. Look at the blood on his muzzle. And he smells like a skunk."

Dick has made a *good argument*:

Spot has blood on his muzzle. Spot smells like a skunk.
Therefore, Spot killed a skunk.

Dick has left out some premises that he knows are as obvious to Zoe as to him:

Spot isn't bleeding.
Skunks aren't able to fight back very well.
Dogs try to kill animals by biting them.
Normally when Spot draws a lot of blood from an animal
    that's smaller than him, he kills it.
Only skunks give off a characteristic skunk odor that
    drenches whoever or whatever is near if they are attacked.

Zoe replies, "Oh, that explains why he's got blood on his muzzle and smells so bad." That is, she takes the same claims and views them as an explanation, a *good explanation*, relative to the same unstated premises:

Spot killed a skunk
explains why Spot has blood on his muzzle and smells like a skunk.

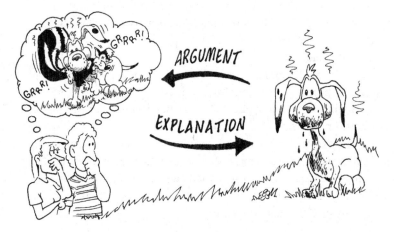

For an explanation "E because of A, B, C, . . ." we can ask what evidence we have for A. Sometimes we can supply all the evidence we need just by reversing the inference. For Zoe's explanation to be good, "Spot killed a skunk" must be plausible, and it is because of the argument Dick gave—they needn't wait until they find the dead skunk.

Explanations and associated arguments   For an inferential explanation:

E because of A, B, C, . . .

the **associated argument** to establish A is:

E, B, C, . . . therefore A

An explanation is **dependent** if one of the premises is not plausible and the associated argument for that premise is not good. An explanation is **independent** if it is not dependent.

If an explanation is dependent, then it lacks evidence for at least one of its premises that cannot be supplied by an associated argument.

Example 25   Spot chases cats because he sees them as something good to eat and because cats are smaller than him.

Analysis   "Cats are smaller than Spot" is clearly true, but "Spot sees cats as something good to eat" is not. The associated argument for it is:

Spot chases cats and cats are smaller than Spot.

Therefore, Spot sees cats as something good to eat.

This is weak. Without more evidence for "Spot sees cats as something good to eat" we shouldn't accept the explanation. The explanation is dependent.

Each premise of an independent explanation is plausible, either because of the associated argument for it or because of independent reasons, such as our knowing that most dogs bark or that Sheila is not a herring. Still, an independent explanation might be bad.

Example 26 Suzy: Why do classes last only 50 minutes instead of an hour?
Maria: Because students need time to get from one class to another.

Analysis The premise is certainly plausible, so the explanation is independent. But it's not good because it's not strong. Why not have classes that last 45 minutes? What follows is only that there should be some time between classes.

Example 27

Analysis Dick offers an explanation:

Light is bent where the water meets the air
explains why the oar appears bent.

The claim doing the explaining here is a generalization: "Light is bent where the water meets the air." Zoe hasn't taken a physics course, so the only reason she has to believe that claim is the associated argument:

The oar appears bent.
Therefore light is bent where the water meets the air.

But this is a weak generalization, needing more examples to convince. So Dick has given a dependent explanation.

Dick, though, has taken a physics course and has studied lots of other examples where light is refracted through various media. So he does have other evidence for his claim that he can offer to Zoe to make this explanation good.

Explanations and predictions

Example 28   Flo: Spot barks. And Wanda's dog Ralph barks.
And Dr. E's dogs Anubis and Juney bark. So all dogs bark.
Barb: Yeah. Let's go over to Maple Street and see if all the dogs there bark, too.
Analysis Flo, who's 5, is generalizing. Her friend Barb wants to test the generalization.

Suppose that *A, B, C, D* are given as inductive evidence for a generalization *G* (some other plausible unstated premises may also be needed, but let's keep those in the background). Then we have that *G* explains *A, B, C, D*.

But if *G* is true, we can see that it follows that some other claims are true, instances of the generalization *G*, say *L, M, N*. If those are true, then *G* would explain them, too (Fido barks, Lady barks, Buddy barks). That is, *G explains A, B, C, D* and *predicts L, M, N*, where *the difference between the explanation and the prediction is that we don't know if the prediction is true* — not that the prediction is about the future.

Suppose we find that *L, M, N* are indeed true. Then the argument *A, B, C, D + L, M, N, therefore G* is a better one for *G* than we had before. At the very least it has more instances of the generalization as premises.

How can more instances of a generalization prove the generalization better? They can if (1) they are from different kinds of situations, that is *A, B, C, D + L, M, N* cover a more representative sample of possible instances of *G* than do just *A, B, C, D*. Typically that's what happens: we deduce claims from *G* for situations we had not previously considered. And (2) because we had not previously considered the kind of instances *L, M, N* of the generalization *G*, we have some confidence that we haven't got *G* by manipulating the data, selecting situations that would establish just this hypothesis.

A good way to test an hypothesis or generalization is to try to falsify it. Trying to falsify a generalization means we are trying to come up with instances of the generalization to test that are as different as we can imagine from the ones we first used in making it. Trying to falsify is just a good way to ensure (1) and (2). We say an experiment *confirms* — to some extent — the (doubtful claims in the) explanation if it shows that a prediction is true.

Comparing explanations

Given two explanations of the same claim, which is better? If one is right and the other wrong, the right one is better. If both are acceptable, we prefer the one that doesn't leave us asking a further "Why?"

Then we prefer the **simpler explanation**: (i) its premises are more plausible, (ii) it is more clearly strong or valid (unstated premises are obvious and more plausible), and (iii) it has fewer steps. All else then being equal, we prefer the stronger explanation.

Example 29 Zoe: How was your walk?
> Dick: Spot ran away again just before we got to the yard.
> Zoe: We better get him. Why does he run away just before
> you come home?
> Dick: It's just his age. He'll outgrow it. All dogs do.

Analysis This sounded like a good explanation until Dick and Zoe found that Spot chased a cat up a telephone pole in the field behind their house. The explanation that Dick gave is not bad. Perhaps in a year or two when Spot is better trained, he won't run away even to chase a cat. But there's a better explanation: Spot ran away because he likes to chase cats and he saw a cat nearby to chase. It's better because it's stronger.

Example 30 Harry: Amazing. Zoe and Dick got into a fight and . . .
> Maria: Zoe's so mad she won't talk to him anymore.
> Tom: How do you know that?
> Suzy: She has ESP!
> Harry: C'mon. She must have heard it from someone else.

Analysis Which is the better explanation? The second because it's more plausible.

Inference to the best explanation

Some people think that if they have an explanation that explains a lot and is the best anyone's offered, then it must be true.

> "It can hardly be supposed that a false theory would explain,
> in so satisfactory a manner as does the theory of natural
> selection, the several large classes of facts above specified
> [the geographical distribution of species, the existence of
> vestigial organs in animals, etc.]. It has recently been
> objected that this is an unsafe method of arguing; but it is

a method used in judging of the common events of life, and
has often been used by the greatest natural philosophers."

Charles Darwin, *On the Origin of Species*

If Darwin is right, why did scientists spend the next hundred years
trying to confirm or disprove the hypotheses of natural selection?
Only now do we believe that a somewhat revised version of Darwin's
hypotheses are true. ***Darwin's mistake*** is to think that if we can easily
deduce a lot of truths from a claim, then the claim must be true. But
that's just arguing backwards (p. 27).

For claims to explain satisfactorily, they have to be plausible —
otherwise the explanation isn't good. Either the associated argument
has to be good or we need independent evidence for the truth of the
claims.

But if it's the best explanation, doesn't that count for it being true?
We don't have accepted criteria for what counts as the best explanation,
and in any case it's only the best explanation we've thought of so far.
Scientists have high hopes for their hypotheses and are motivated to
investigate them if they appear to provide a better explanation than
current theories. But the scientific community quickly corrects anyone
who thinks that just making an hypothesis that explains a lot establishes
that it's true. The ***fallacy of inference to the best explanation*** is to
claim that because some claims constitute the best explanation we have,
they're therefore true. Finding an explanation that's better than all
others we have gives us at most good motive to investigate whether the
claims doing the explaining are true.

> *This is the best explanation we have*
> *= this is a good hypothesis to investigate.*

Example 31 Tom: The AIDS epidemic was started by the CIA. They
wanted to get rid of homosexuals and blacks, and they targeted those
groups with their new disease. They started their testing in Africa in
order to keep it hidden from people here. The government once again
tried to destroy people they don't like.

Analysis That the AIDS epidemic was started by the CIA would
explain a lot. But that's no reason to believe it's true. The only evidence we have for it is the associated argument, which is weak.

Every conspiracy theory depends on Darwin's mistake to make
us believe it. If someone says that the explanation isn't the best one

because its premises aren't plausible, then they've already accepted that we have to have independent evidence for the explanation being true.

Example 32 Me: Why do I have such pain in my back? It doesn't feel like a muscle cramp or a pinched nerve.

Physician: A kidney stone would explain the pain. Kidney stones give that kind of pain, and it's in the right place for that.

Analysis When I went into the emergency room one night, the doctor gave me this explanation. It would have been a good one if he'd had good reason to believe "You have a kidney stone." But at that point the only reason he had was the associated argument, and that wasn't strong. Still, it was the best explanation he had.

So the doctor made predictions, reasoning by hypotheses: "A kidney stone would show up on an X-ray," "You'd have an elevated white blood cell count," "You would have blood in your urine." He tested each of these and found them false. He then reasoned by reducing to the absurd that if the explanation were true, these would very likely be true; they are false; therefore, the explanation is very likely false.

Nothing else was found, so by process of elimination the doctor concluded that I had a severe sprain or strain, for which exercise and education were the only remedy.

If the doctor had believed "You have a kidney stone" just because that was the best available explanation, there would have been no point in doing tests. And then I would have undergone needless treatment or even surgery. Finding an explanation that's better than all others does not justify belief, but is only a good motive to investigate whether it's true.

Teleological explanations

One day while cleaning out the small pond in my backyard I asked myself, "Why is there a filter on this wet-dry vacuum?" The vacuum had a sponge-like filter, but the vacuum sucked up water a lot faster without it. I wondered if I could remove the filter.

I wanted to know the function of the filter. A causal explanation could be given starting with how someone designed the vacuum with the filter, invoking what that person thought would be its function. But most of that explanation would be beside the point. I didn't want to know why it's true that there's a filter on the vacuum, even though the truth of that claim is assumed in the question. I wanted to know the function of the filter. Some explanations should answer not "Why is this true?" but "What does this do?" or "Why would he or she do that?"

Teleological explanations A *teleological explanation* is an explanation that invokes goals or functions or uses claims that can come true only after the claim(s) doing the explaining.

Example 33 Why is the missile going off in that direction?
   Because it wants to in order to hit that plane.
   Analysis It's a bad anthropomorphism to ascribe goals to a missile: People, not missiles, have goals. We should replace this teleological explanation with an inferential one: "The missile has been designed to go in the direction of the nearest source of heat comparable to the heat generated by a jet engine. The plane over there in that direction has a jet engine producing that kind of heat." Often a teleological explanation is offered when an inferential one should be used.

There are serious problems with teleological explanations. If an explanation uses claims that can be true only after the claim being explained becomes true, then it can't be causal (the cause has to precede the effect) and it would seem that the future is somehow affecting the past.

Moreover, a request for a teleological explanation assumes that the object has a function or that the person or thing has a goal or motive. That's part of what's being assumed. But often there is simply no motive, no function, no goal, or at least none we can discern. The right response, as to a loaded question, is to ask why we should believe there is a function or motive. The *teleological fallacy* is to assume that because something occurs in nature, it must have a purpose.

Example 34 Dick (picking his nose): Why do humans get snot in their nose that dries up and has to be picked away? I can't understand what good it does.
   Zoe: What makes you think there's a purpose? Can't some things just be? Maybe it just developed along with everything else.
   Analysis Zoe correctly points out Dick's teleological fallacy.

A bigger problem with teleological explanations is that we don't have criteria for what counts as a good one. That's because we don't have a clear idea how to judge what counts as the function of something.

At best, we can say that for a good teleological explanation *E because of A, B, C,* . . . :

* *E* should be highly plausible.
* The explanation should answer the right question.
* It shouldn't be circular.
* It shouldn't ascribe motives, beliefs, or goals to something that doesn't have those.

Example 35 Why do we dream? Dreams serve as wish-fulfillments to prevent interruption of sleep, which is essential to good health.

Analysis According to Freudians, this is a good explanation. And it's teleological. There seems to be no way to construe it as inferential, at least not in accord with the rest of Freud's theory.

Opinion now divides: either this is an example of a good explanation that is truly teleological, or this example shows that Freudian theories of the unconscious are no good because they yield only teleological explanations.

Example 36 Why does the blood circulate through the body?

(1) Because the heart pumps the blood through the arteries.

(2) In order to bring oxygen to every part of the body tissue.

Analysis The first explanation is a good causal one, if it answers the right question. The second is a good teleological one, if it answers the right question.

Example 37 Dick: There. See! The sign for the western menswear store: "Real Men Don't Browse."

Zoe: Yeah. So?

Dick: It's true. We're genetically programmed that way. Long ago when humans were evolving, men were stronger and went out to hunt. The women gathered fruit and berries. When we're hunting we take the first thing we can get, maybe just a rabbit, even if we're looking for a mastodon, 'cause we don't know if there'll be another chance that day. When women went out to get berries there were lots of choices, so they'd pick the best and discard ones that weren't quite so good. That's why women like to shop and men just go in and buy the first thing that looks good and then leave.

Analysis This explains why men don't browse and women do. But it doesn't explain it well. It's an example of Darwin's mistake, an evolutionary *just-so story*, like Kipling's "How the Leopard Got Its Spots."

22 Models and Theories

What is a model? How do we determine if a model is good? How can we modify a model in the light of new evidence?

Examples of models and theories

Example 1   A map of Minersville, Utah—reasoning by analogy

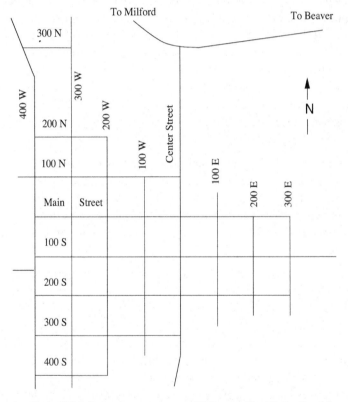

Analysis   This is an accurate map of Minersville, Utah. Looking at it we can see that the streets are evenly spaced. For example, there is the same distance between 100 N and 200 N as between 100 E and 200 E. The last street to the east is 300 E. There is no paved road going north beyond Main Street on 200 E.

That is, from this map we can deduce claims about Minersville, even if we've never been there. But there is much we can't deduce:

Are there hills in Minersville? Are there lots of trees? How wide are the streets? How far apart are the streets? Where are there houses? The map is accurate for what it pays attention to: the relative location and orientation of streets. But it tells us nothing about what it ignores.

The differences between the map and Minersville aren't important when we infer that the north end of 200 W is at 200 N. In contrast, a scale model of a city or a mountain abstracts less from the actual terrain: height and perhaps placement of rivers and trees are there. The map of Minersville *abstracts more* from the actual terrain than a scale model of the city would—that is, *it ignores more*.

To use this model is to reason by analogy: we can draw conclusions when appropriate similarities are invoked and the differences don't matter. The general principle, in this example, is not stated explicitly. The discussion above suggests how we might formulate one, but it hardly seems worth the effort. We can "see" when someone has used a map well or badly.

Example 2 Models of the solar system
Here is a sketch of the model of the universe the Egyptian astronomer Ptolemy proposed in the second century A.D.

Ptolemy's model

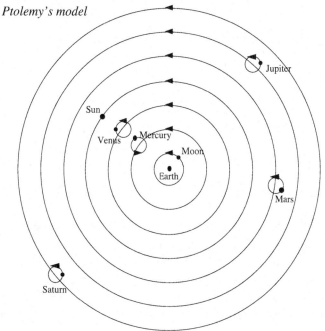

It's meant to show the relative positions of the planets, sun, and moon, and the ways they move. We can't deduce anything about, say, the size of the planets, the distances between them, nor the speeds at which they move, because this model ignores those. According to this model, the moon, sun, and each of the planets revolves around the Earth in a circular orbit, all moving in the same direction. Along that orbit, each planet also revolves in a smaller circle, called an "epicycle." The sun, Earth, and Venus are always supposed to be in a line as shown in the picture.

Ptolemy made a lot more claims about the planets, Earth, and sun that were to be used in making predictions, but for our purposes this sketch will do.

Ptolemy's model accorded pretty well with observations of the movements of the planets and was the generally accepted way to understand the universe for many centuries. But in 1543 the Polish astronomer Copernicus published a book with a different model of the universe.

Copernicus' model

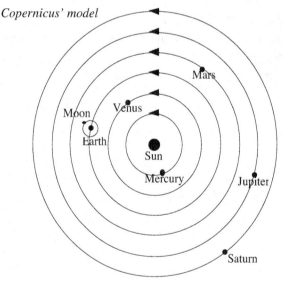

This sketch, too, abstracts a lot from what is being modeled. The sun is shown to be larger than the planets, but that's all we see about their relative sizes. We can't tell from the picture whether the orbits are all on the same plane or on different planes. We do see that the planets all revolve in the same direction and that the Earth, sun, and Venus do not always stay lined up.

Ptolemy accounted for the motion of the sun, planets, and stars in the sky by saying they revolved around the Earth every 24 hours. Copernicus accounted for those motions by saying that the Earth revolved around its own axis every 24 hours. How could someone in the late 16th century decide between these two models? Both were in accord with the observations that had been made.

In the early 1600s the telescope was invented, and in 1610 Galileo built his own telescope with a magnification of about 33 times, using it to study the skies. One of his students suggested an experiment that might distinguish between the Ptolemaic and Copernican models. Venus was too far from the Earth to be seen as anything other than a spot of light. But according to Ptolemy's model, viewed from the Earth, at most only a small crescent-shaped part of Venus will be illuminated by the sun. From Copernicus' model, however, we can deduce that from the Earth, Venus should go through all the phases of illumination, just like the moon: full, half, crescent, dark, and back again. Galileo looked at Venus through his telescope for a period of time and saw that it exhibited all phases of illumination, and this he took to be proof that Copernicus' model was correct.

Not a lot of other people were convinced, however. Telescopes were rare and not very reliable: they introduced optical illusions, such as halos, from the imperfections in the glass and the mounting. Why should astronomers have trusted Galileo's observations?

It was more due to Newton that something like Copernicus' model of the universe was finally accepted. Newton deduced from his laws of motion that the orbits of the Earth, sun, and the planets would have to be ellipses, not circles. And the distances between them would have to be much greater than supposed. Using Newton's laws, Edmond Halley predicted correctly the return of a comet that had been observed in 1682. Telescopes were better, with fewer optical illusions, and they were common enough that most astronomers could use one, so better and better observations of the planets and stars could be made. Those observations could be deduced from the Copernican-Newtonian model, while new epicycles had to be invented to account for them in the Ptolemaic model.

Note that each model is supposed to be similar to the universe in only a few respects, ones that would have an effect on how we could see the objects in the universe from the Earth. Differences, such as whether Venus is rocky or gaseous, are not supposed to matter for those observations. If the model is correct, then reasoning by analogy — very precise analogy — certain claims can be deduced.

Example 3   The kinetic theory of gases — getting true predictions
            doesn't mean the model is true

"This theory is based on the following postulates, or assumptions.

1. Gases are composed of a large number of particles that behave like hard, spherical objects in a state of constant, random motion.

2. The particles move in a straight line until they collide with another particle or the walls of the container.

3. The particles are much smaller than the distance between the particles. Most of the volume of a gas is therefore empty space.

4. There is no force of attraction between gas particles or between the particles and the walls of the container.

5. Collisions between gas particles or collisions with the walls of the container are perfectly elastic. Energy can be transferred from one particle to another during a collision, but the total kinetic energy of the particles after the collision is the same as it was before the collision.

6. The average kinetic energy of a collection of gas particles depends on the temperature of the gas and nothing else."

J. Spencer, G. Bodner, and L. Rickard, Chemistry

Analysis Here is a picture of what is supposed to be going on in a gas in a closed container. The molecules of gas are represented as dots, as if they were hard spherical balls. The length of the line emanating

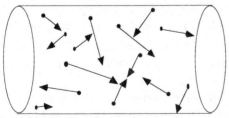

from a particle models the particle's speed; the arrow models the direction in which the particle is moving. The kinetic energy of a particle is defined in terms of its mass and velocity: kinetic energy = .5 mass $\times$ velocity^2. The model defines what is meant for a collision to be elastic. In contrast, here is a picture of what happens in an inelastic collision between a rubber ball and the floor.

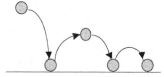

Each time the ball hits the ground, some of its kinetic energy is lost either through being transferred to the floor or in compressing the ball.

What are we to make of these assumptions? Some are false: molecules of gas are not generally spherical and are certainly not solid; the collisions between molecules and the walls of a container or each other are not perfectly elastic; there is some gravitational attraction between the particles and each other and also with the container. How can we use false claims in a model?

The model proceeds by abstraction, much like in analogy: to the extent that we can ignore how molecules of gases are not spherical, and ignore physical attraction between molecules, and ignore . . . we can draw conclusions that may be applicable to actual gases. To the extent that the differences between actual gases and the abstractions don't matter, we can draw conclusions. But how can we tell if the differences matter?

The model suggests that the pressure of a gas results from the collisions between the gas particles and the walls of the container. So if the container is made smaller for the same amount of gas, the pressure should increase; and if the container is made larger, the pressure should be less. So the pressure should be proportional to the inverse of the volume of the gas. That is, the model suggests a claim about the relationship of pressure to volume in a gas. Experiments can be performed, varying the pressure or volume, and they are close to being in accord with that claim.

Other laws are suggested by the model: Pressure is proportional to the temperature of the gas, where the temperature is taken to be the average kinetic energy of the gas. The volume of the gas should be proportional to the temperature. The amount of gas should be proportional to the pressure. All of these are confirmed by experiment.

Those experiments confirming predictions from the model do not mean the model is more accurate than we thought. Collisions still aren't really elastic; molecules aren't really hard spherical balls. The kinetic theory of gases is a model that is useful, as with any analogy, when the differences don't matter.

Example 4   The acceleration of falling objects—
         an equation can be a model
Galileo argued that falling objects accelerate as they fall: they begin
falling slowly and fall faster and faster the farther they fall. He didn't
need any mathematics to show that. He just noted that a heavy stone
dropped from 6 feet will drive a stake into the ground much farther than
if it were dropped from 6 inches.

He also said that the reason a feather falls more slowly than an iron
ball when dropped is because of the resistance of air. He argued that at
a given location on the Earth and in the absence of air resistance, all
objects should fall with the same acceleration. He claimed that the
distance traveled by a falling object is proportional to the square of the
time it travels. Today, from many measurements, the equation is given
by:

(*) $d = \frac{1}{2}\, 9.82 \text{ meters/sec}^2 \cdot t^2$ where t is time in seconds

Analysis   The equation (*) is a model by abstraction: we ignore
air resistance and the shape of the object, considering only the object's
mass and center of gravity. If the differences don't matter, then a
calculation from the equation, which is really a deduction, will hold.
But often the differences do matter. Air resistance can slow down an
object: if you drop a cat from an airplane, it will spread out its legs and
reach a maximum velocity when the force of the air resistance equals
the force of acceleration.

With this model there is no visual representation of that part of
experience that is being described. There is no point-to-point concep-
tual comparison, nor are we modeling a static situation. The model is
couched in the language of mathematics; equations can be models, too.

Example 5   Newton's laws of motion and Einstein's theory of
         relativity—how a false theory can be used
Newton's laws of motion are taught in every elementary physics course
and are used daily by physicists. Yet modern physics has replaced
Newton's theories with Einstein's and quantum mechanics. Newton's
laws, physicists tell us, are false.

But can't we say that Newton's laws are correct relative to the
quality of measurements involved, even though Newton's laws can't be
derived from quantum mechanics? Or perhaps they can if a premise is
added that we ignore certain small effects. Yet how is that part of a
theory?

A theory is a schematic representation of some part of the world. We draw conclusions from the representation (we calculate or deduce). The conclusion is said to apply to the world. The reasoning is legitimate so long as the differences between the representation and what is being represented don't matter. Newton's laws of motion are "just like" how moderately large objects interact at moderately low speeds; we can use those laws to make calculations so long as the differences don't matter. Some of the assumptions of that theory are used as conditions to tell us when the theory is meant to be applied.

Example 6   Ether as the medium of light waves — a prediction can show that an assumption of a theory is false
In the 19th century light was understood as waves. In analogy with waves in water or sound waves in the air, a medium was postulated for the propagation of light waves: the ether. Using that assumption, many predictions were made about the path and speed of light in terms of its wave behavior. Attempts then were made to isolate or verify the existence of an ether. The experiments of Michaelson and Morley showed those predictions were false. When a better theory was postulated by Einstein, one which assumed no ether and gave as good or better predictions in all cases where the ether assumption did, the theory of ether was abandoned.

Example 7   Euclidean plane geometry—a model that can't be true
Euclidean plane geometry speaks of points and lines: a point is location without dimension, a line is extension without breadth. No such objects exist in our experience. But Euclidean geometry is remarkably useful in measuring and calculating distances and positions in our daily lives.

Points are abstractions of very small dots made by a pencil or other implement. Lines are abstractions of physical lines, either drawn or sighted. So long as the differences don't matter, that is, so long as the size of the points and the lines are very small relative to what is being measured or plotted, we can deduce conclusions that are true.

No one asks (anymore) whether the axioms of Euclidean geometry are true. Rather, when the differences don't matter, we can calculate and predict using Euclidean geometry. When the differences do matter, as in calculating paths of airplanes circling the globe, Euclidean plane geometry does not apply, and another model, geometry for spherical surfaces, is invoked.

Euclidean geometry is a deductive theory: a conclusion drawn from the axioms is accepted only if the inference is valid. It is a purely mathematical theory, which taken as mathematics would appear to have no application since the objects of which it speaks do not exist. But taken as a model it has applications in the usual way, arguing by analogy where the differences don't matter.

Models, analogies, and abstraction

We've seen models of static situations (the map) and of processes (acceleration of falling objects). We've seen examples of models that are entirely visual and of models formulated entirely in terms of mathematical equations. We've seen models in which the assumptions of the model are entirely implicit (the map), and we have seen models in which the assumptions are quite explicit (Newton's laws of motion).

In all the examples either the reasoning is clearly reasoning by analogy or can be seen to proceed by abstraction much as in reasoning by analogy. We do not ask whether the assumptions of a theory or model are true, even if that was the intention of the person who created the theory. Rather, we ask whether we can use it in the given situation: Do the similarities that are being invoked hold and do the differences not matter? Even in the case of Newton's laws of motion, where it would seem that what is at stake is whether the assumptions are true, we continue to use the model when we know that the assumptions are false in those cases where, as in any analogy, the differences don't matter. In only one example (the ether) did it seem that what was at issue was whether a particular assumption of the theory was actually true of the world.

The assumptions of theories in science are false when we consider them as representing all aspects of some particular part of our experience. The key claim in every analogy is false in the same way. When we say that one side of an analogy is "just like" the other, that's false. What is true is that they are "like" one another in some key respects that allow us to deduce claims for the one from deducing claims for the other.

The term *model* is typically applied to what can be visualized or made concrete, while *theory* seems to be used for examples that are fairly formal with explicitly stated assumptions. But in many cases it is as appropriate to call an example a theory as to call it a model, and there seems to be no definite distinction between those terms.

Confirming a theory

From theories we can make predictions. When a prediction turns out to be true, we say it *confirms* (to some degree) the theory. This is not the same as confirming an explanation, for it rarely makes sense to say that the claims that make up a theory—the assumptions of the theory—are true or false.

We cannot say that verifications of the relation of pressure, temperature, and volume in a gas confirm that molecules are hard little balls and that all collisions are completely elastic. Nor does fitting a carpenter's square exactly into a wooden triangle that is 50cm x 40cm x 30cm confirm the theorem of Pythagoras. Nor can we say that finding a tree at the corner of 100 W and 100 S in Minersville disconfirms the model given by the map. The map wasn't meant to give any information about trees, so it doesn't matter that it shows no tree there.

Except in rare instances where we think (usually temporarily) that we have hit upon a truth of the universe to use as an assumption in a theory, we do not think that the assumptions of a theory are true or false. We can only say of a theory such as Euclidean plane geometry or the kinetic theory of gases whether it is *applicable* in a particular situation we are investigating.

To say that a theory is applicable is to say that, though there are differences between the world and what the assumptions of the theory state, those differences don't matter for the conclusions we wish to draw. Often we can decide if a theory is applicable only by attempting to apply it. We use the theory to draw conclusions in particular cases, claiming that the differences don't matter. If the conclusions—the predictions—turn out to be true (enough), then we have some confidence that we are right. If a prediction turns out false, then the model is not applicable there. We do not say that Euclidean plane geometry is false because it cannot be used to calculate the path of an airplane on the globe; we say that Euclidean plane geometry is inapplicable for calculating on globes.

When we make predictions and they are true, we confirm a range of application of a model. When we make predictions and they are false, we disconfirm a range of application—that is, we find limits for the range of application of a model. More information about where the model can be applied and where it cannot be applied may lead, often with great effort, to our describing more precisely the range of application of a model. In that case, the claims describing the range of application can be added to the theory. We often use mathematics as a language to make this art of analogy precise. But in many cases it is difficult to state precisely the range of application. Reasoning using models is reasoning by analogy, which is likely to require judgment.

Sometimes it's said that a theory is valid, or is true, or that a theory holds. These are just different ways to assert that a particular situation or class of situations to which we wish to apply a theory is within the range of that theory.

Good theories, and modifying theories in the light of new evidence

We've seen that the criteria for whether one theory is good or better than another cannot in general include whether the assumptions of the theory are true or "realistic." So besides getting true predictions, what criteria can we use to evaluate theories? Consider what we do when we discover that a prediction made from a theory is false.

When Newton's laws of motion result in inaccurate predictions for very small objects, we note that the theory had been assumed true for all sizes and speeds of objects and then restrict the range of application. But when the theory of the ether resulted in false predictions, no modification was made to the theory, for none could be made. That theory did not abstract from experience, ignoring some aspects of situations under consideration, but postulated something in addition to our experience, which we were able to show did not exist. The theory was completely abandoned.

If a theory has been made by abstraction—that is, many aspects of our experience are ignored and only a few are considered significant— then tracing back along that *path of abstraction* we can try to distin- guish what difference there is between our model and our experience that matters. What have we ignored that cannot in this situation be ignored? If we cannot state generally what the difference is that matters, then at best the false prediction sets some limit on the range of applicability of the model or theory. We cannot use the theory here —where "here" means this situation or ones that we can see are very similar.

But our goal will be to state precisely the difference that matters and try to factor it into our theory. We try to devise a complication of our theory in which that aspect of our experience is taken into account. As with Einstein's improvement of Newton's laws, we get a better theory that is more widely applicable and that explains why the old theory worked as well as it did and why it failed in the ways it failed. We improve the map: by adding more assumptions, we can pay attention to more in our experience, and that accounts for the differences between the theories.

True predictions are never enough to justify a theory. Indeed, we do not "justify" a theory, nor show that it is "valid." What we do in the process of testing predictions is show how and where the theory can be applied. And for us to have confidence in that, either i. we must show that the claims in the theory are true, or ii. show in what situations the differences between what is being modeled and the abstraction of it in the theory do not matter. True (enough) predictions help in that. But equally crucial is our ability to trace the path of abstraction so that we can see what has been ignored in our reasoning and why true predictions serve to justify our ignoring those aspects of experience. Without that clear path of abstraction, all we can do is try to prove that the claims in the theory are actually true.

Sometimes we are confronted with two theories that both yield good predictions for a class of situations and both of which have a clear path of abstraction. In that case, we say that one theory is **better** than another if (1) its assumptions are simpler; (2) it yields clearer derivations of the claims it is meant to explain; (3) it has a wider range of application; and (4) it yields better explanations of the archetypal claims it is meant to explain.

Example 8 "Consider the density of leaves around a tree. I suggest the hypothesis that the leaves are positioned as if each leaf deliberately sought to maximize the amount of sunlight it receives, given the position of its neighbors, as if it knew the physical laws determining the amount of sunlight that would be received in various positions and could move rapidly or instantaneously from any one position to any other desired and unoccupied position. Now some of the more obvious implications of this hypothesis are clearly consistent with experience: for example, leaves are in general denser on the south than on the north side of trees but, as the hypothesis implies, less so or not at all on the northern slope of a hill or when the south side of the trees is shaded in some other way. Is the hypothesis rendered unacceptable or invalid because, so far as we know, leaves do not 'deliberate' or consciously 'seek,' have not been to school and learned the relevant laws of science or the mathematics to calculate the 'optimum' position, and cannot move from position to position? Clearly, none of these contradictions of the hypothesis is vitally relevant; the phenomena involved are not within the 'class of phenomena the hypothesis is designed to explain'; the hypothesis does not assert that leaves do these things but only that their density is the same *as if* they did. Despite the apparent falsity of the 'assumptions' of the hypothesis, it has great plausibility because of the conformity of its implications with observation. We are inclined to 'explain' its validity on the ground that sunlight contributes to the growth of leaves and that hence leaves will grow denser or more putative leaves survive where there is more sun, so the result achieved by purely passive adaptation to external circumstances is the same as the result that would be achieved by deliberate accommodation to them. This alternative hypothesis is more attractive than the constructed hypothesis not because its 'assumptions' are more 'realistic' but rather because it is part of a more general theory that applies to a wider variety of phenomena, of which the position of leaves around a tree is a special case, has more implications capable of being contradicted, and has failed to be contradicted under a wider variety of circumstances."

Milton Friedman, "The Methodology of Positive Economics"

Analysis Friedman's hypothesis about leaves seeking to maxi-mize the amount of sunlight they receive cannot be used for reasoning by analogy or abstraction. It does not begin by either (a) looking at a real situation and comparing it to the growth of leaves, allowing us to distinguish what are the similarities and what are the differences, or (b) abstracting from experience to state what are the points of similarity that are supposed to hold, ignoring all else.

Rather, what he has posited is not an abstraction, but the addition of properties to a given situation. We are asked to suppose that leaves behave anthropomorphically with the skills of a terrific calculator. And then we are asked to ignore that as well. This doesn't make sense as a method of reasoning: why should we have confidence that predictions made from such a hypothesis will be accurate? That some of the pre-dictions turn out to be accurate cannot be enough, any more than they are in astrology. We need to know why they turn out accurate in order to have confidence in the theory or model.

The alternative hypothesis of passive adaptation that Friedman presents is better, but not for the reasons he gives; rather, it is better for the reason he says is not meaningful. Namely, we have better reason to accept the alternative hypothesis precisely because we can see that in this case it is reasonable to believe it is true. No clearly false assumption incapable of fitting into reasoning by analogy or abstraction has been made.

Evaluating Risk, Making Decisions

23 Evaluating Risk

Weighing risk

A *risk* is a possibility in the future that we deem bad and that would be a consequence of some action we take or don't take.

Example 1 Dick likes to let Spot run free when they go for a walk, and a risk associated with doing so is that Spot could chase a cat into traffic and get hit by a car. Perhaps Dick never thought of that risk, in which case he's not very thoughtful about what he does. But he has considered it, and he's decided that letting Spot run free is *worth the risk*: the good that might come from it outweighs, by more than a little, the bad that might come from it. If Dick doesn't let Spot run free, Spot never gets a chance to get any real exercise, and Dick doesn't think it's very likely that Spot will run out into traffic. He's a good dog.

Tom disagrees. He's seen Spot go crazy around cats, and he knows that's not the only reason Spot might run into traffic. Tom knows, too, that there are other ways for Spot to get exercise: Dick could lead him on a leash while Dick is bicycling or jogging, or Dick could take Spot to the countryside a few times a week to run free. Tom is weighing the likelihood of the risk differently and is seeing more choices for accomplishing the good that Dick wants.

Evaluating risk To evaluate a course of action we need to:
- Weigh how likely the risk is.
- Weigh how likely a good outcome is.
- Weigh how much we want to avoid the risk versus how much we want the possible good outcome that might come from our action.
- Weigh how hard it would be to accomplish the good by other means.

Only rarely can we put numbers to these evaluations. How much we want to avoid a risk or get a good outcome is a subjective or at best an intersubjective evaluation.

Example 2 Harry is a hard-working student and he'd really like to make $5,000. He knows a stock that he's sure is undervalued that he could buy for $5,000 and make at least that much in a few months. But he doesn't want to make money so much that he's willing to risk losing almost all of his savings. In comparison, for Bill Gates losing $5,000 would be a negligible risk, while making $5,000 would be OK but hardly worth the trouble.

How hard it would be to accomplish the good by other means usually depends on subjective criteria, too.

Example 3 It would be easy for Tom to jog or bicycle with Spot at his side on a leash because Tom plays football and enjoys exercising. But for Dick, who doesn't get much exercise, that would be hard to do. It would be easier for Dick to ask Tom to take Spot for a run, since it was Tom's idea anyway.

Example 4 Dick: Zoe! Did you hear the weather report? There's a 40% chance of a tornado in the next hour.

Zoe: We better open some windows and get into the basement. Grab a flashlight and the radio.

Analysis A tornado is certainly bad, but it's not a risk, for it's not a consequence of something we do or don't do. Insurance companies call a tornado a "natural disaster" or an "act of God." What is a risk is what might happen to Dick and Zoe if they don't go into the basement.

Example 5 The Union of Scientists Worried about Nuclear War estimates there's a 30% chance of a nuclear war somewhere in the world this year.

Analysis Many people view the possibility of a nuclear war as if it were a natural disaster, completely out of their control. But whether it happens or not depends on what we do, all of us from a person who doesn't vote, to a diplomat, to the President. To think of nuclear war as an act of God is to imagine ourselves powerless in the face of unseen forces, when the forces are really of our own creation.

How likely is the risk?

Sometimes we have a pretty good sense of how likely the risk is. Tom thinks it's not at all unlikely that Spot will run off into traffic if Spot's not on a leash, and his evaluation seems more accurate than Dick's wishful thinking. In comparison, Harry can't even begin to guess how likely it is he could lose most of his money if he invests in that stock.

These evaluations of likelihood of a risk or of a good outcome are evaluations of how likely possibilities are, which we've been doing all along in evaluating arguments, causal claims, and explanations. We evaluate the strength of the inference from premises describing the situation now whose conclusion is the outcome we're worried about. For example, Tom thinks that the inference from "Dick lets Spot run free on a walk" to "Spot runs into traffic and gets hit by a car" is much stronger than Dick does. On the other hand, Harry doesn't have a clue how strong the inference is from "I invest $5,000 in a stock, and the stock appears to me to be undervalued" to "I lose most of my money."

Sometimes, however, we can make a very good estimate of how likely the risk is.

Example 6 Dick: I'll bet you $5 on the next flip of this coin.
Tom: You're on.
Analysis Tom knows that the likelihood he'll lose $5 is 50% and the likelihood he'll win $5 is 50%.

Example 7 Zoe: You're buying a lottery ticket?
Dick: Sure. Why not?
Zoe: Can't you do the math? There's less than one chance in 200 million that you'll win.
Analysis Dick's chances of losing $1 is almost certain; his chance of winning is only a tiny bit better than not buying a ticket at all. Yet he still buys a ticket. He reckons that the possible good that can come from buying a ticket, which includes a week of daydreams, out-weighs what he considers the minimal loss. He also consoles himself that the money he loses will go to college scholarships, though very little of it does. Others, however, would classify as a bad outcome Dick passing his time daydreaming what he'd do with millions of dollars.

Example 8 Dick: You thought I was crazy to buy a lottery ticket, yet you just bought six of them.
Zoe: Yes, but the jackpot is up to $460 million!
Analysis Buying six tickets gives Zoe six times more chances to win than buying one ticket—but still that's only a miniscule chance. Either knowing the numbers associated with the risk doesn't matter to Zoe, or else she considers the outcome of winning so much money ver-sus the usual 25 million dollars many million times more good to her. It isn't that the lottery is a tax on people who don't understand mathe-matics; rather, people motivated by greed are willing to weigh a posi-tive outcome much more than the risk, if the risk doesn't cost much.

Examples of mistakes in evaluating risk

Example 9   Tom: I'm thinking of joining the Army.
   Zoe: That's really dangerous.
   Tom: No it isn't. The death rate for soldiers in Iraq is lower than the death rate in New York City.
   Analysis Tom is comparing apples and oranges. The death rate for people in New York includes all the old people there, all the sick people, all the homeless. To evaluate the risk, Tom should compare the death rate for healthy New Yorkers who are of military age to that for soldiers in Iraq, and then factor in the rate of serious injury, particularly brain injury, for those two groups.

Example 10   Zoe: Look in the newspaper! There was a plane crash in New York and all 168 people on board died. When we go to visit my mom this year, let's drive and not fly. It'll be a lot safer.
   Analysis This is a case of selective attention. The likelihood of being killed in an auto accident per miles traveled is much higher than for flying, it's just that we don't read about so many people dying all at once. There's also the illusion that because we're driving we can control the risk more than in a plane. But an airplane pilot is much better trained and careful at flying than most of us are at driving.

Example 11   Dick: I read that most car accidents occur within three miles of the driver's home. So if we do drive to see your mom, we don't need to be so careful once we're out of town.
   Analysis Of course most car accidents happen within three miles of the driver's home because that's where people drive the most. It's the rate of accidents that indicates the likelihood of the risk.

Example 12   Tom: I just read in the paper that a third of all avalanche deaths in the U.S. occur in Colorado. I'm definitely not going skiing there this year.
   Harry: Great. Try Florida. They didn't have any avalanche deaths there last year.
   Tom: Duh. They don't have any snow.
   Harry: And Colorado is where people are skiing.
   Analysis Tom's confused. He thinks that the important figure for determining his risk while skiing is the number of deaths per state. Harry shows he's wrong. The important figure to know is how many of those who are skiing were killed in an avalanche.

Example 13 Zoe: You drink the tap water here? Don't you know that it's high in arsenic?

Zoe's mother: It hasn't killed me yet.

Analysis Sure she hasn't died yet: too much arsenic in drinking water won't kill you instantly. Its effect is cumulative: if you drink too much over a long period of time you'll get cancer. Zoe's mother is confusing short-term risk with long-term risk.

Example 14 Lee's grandfather: It's been 30 years since I smoked and I still crave a cigarette. I'm 75 now, and that's the life expectancy of a man. So I might as well start smoking—the worst it can do is kill me.

Analysis Lee's grandfather's fatalism is misplaced. That the life expectancy of a man is 75 means a newborn male will, on average, live that long. But if a man is 75, on average he'll live another 10.5 years. And by saying that the worst it can do is kill him he's choosing to ignore the other risks of smoking: it'll make it harder for him to breathe and make his last years much worse even if it doesn't kill him.

Evaluating risk in health-care decisions

In decisions about our health we're often given figures that seem to make clear exactly what the likelihood of the risk is versus a good outcome. But how we understand those numbers and whether those numbers really do tell us what we need to know are crucial for us to make good decisions.

Example 15 Suppose major league baseball has a test to detect steroid use which correctly identifies users 95% of the time and misidentifies nonusers 5% of the time. 400 players are tested, and Ralph's test comes out positive. Should he be suspended?

Analysis Sure, we think, there's a 95% chance he's using steroids. But that's not right. Previous testing showed that only about 5 percent of baseball players use steroids. So of the 400 players who were tested:

20 (5%) are users.

380 are not users.

Of the 20 users, 19 (95%) are likely correctly identified as users.

Of the 380 nonusers, 19 (5%) are likely incorrectly identified
   as users.

So there's only a 50% chance that Ralph, one of the 38, tested positive because he uses steroids. It isn't just the accuracy of the test, but also what proportion of players who use steroids that determines how likely a positive test is right. That's why they repeat the test before suspending someone. After a second test for these 38, of the 19 false positives at most we'd expect one false positive again (5%), while of the ones that were users, 17 would likely test as being users (95%). After a third test, there's almost no chance that a person who isn't a user will be classified as one, and it's almost certain that someone who tests positive is a user. *To evaluate the outcome of a test we need to consider not just how reliable the test is but also what we know about what proportion of the population actually is or does what's tested.*

Example 16   "Doctors report that, in treating men with cancer that has started to spread beyond the prostate: Survival is significantly better if radiation is added to standard hormone treatments.

The new study assigned 1,200 men to get hormones plus radiation or hormones alone. After seven years, 74 percent of the men receiving both treatments were alive versus 66 percent of the others.

Those on both treatments lived an average of six months longer than those given just hormones."   *Albuquerque Journal*, 6/7/2010

Analysis   Suzy's grandfather has prostate cancer that's begun to spread. After she read this she urged him to get radiation therapy along with hormones. She told him, "You'll live six months longer!"

Her grandfather pointed out that men live six months longer on average, which is no guarantee he'll live that much longer. And he pointed out that the report doesn't talk about the bad effects of taking radiation therapy, which could make a man wish he hadn't lived six months more. Considering only the possible good is no basis for making a decision.

Example 17   Zoe: I can't believe you're taking St. John's wort!

Zoe's mom: They say it's good for depression. It helped my friend Sally. And besides, it can't hurt. They sell it at the natural foods store.

Analysis   Evaluating risk is evaluating reasoning. Who's "they"? Is Zoe's mom just repeating what she heard somewhere? That it helped her friend Sally could be coincidence, and even if it weren't, at best it would be anecdotal evidence. And the idea that if something is sold at a natural foods store then it can't hurt you is nonsense. With a little searching on the Web Zoe can show her mom that researchers at Duke

University Clinical Research Institute have found that St. John's wort can interfere with other medications. And even if it were harmless, taking it instead of seeking professional help for depression can be harmful. Saying "Oh well, it can't hurt" is just a way to avoid thinking seriously about bad consequences.

Example 18 "A vaccine for shingles was licensed in 2006. In clinical trials, the vaccine prevented shingles in about half of people 60 years of age and older."   "Shingles vaccine: What you need to know"
Website of the Centers for Disease Control, 9/12/2006

Analysis This is an authoritative source and seems to make clear that if someone 60 or older takes the vaccine, they'll reduce the chance of getting shingles by 50%. But elsewhere on the website of the Centers for Disease Control it says that 1 in 3 people that age develop shingles in their lifetime. So there's a $2/3$ chance someone that age won't develop shingles anyway. The only reading of the CDC's statement that makes any sense is that one-half as many people in the trial got shingles as would have been expected without taking the vaccine. That means that if you're 60 or older, your chance of getting shingles is $1/3$ without the vaccine, and $1/6$ with the vaccine—if you don't know in advance whether you are one of those who is not likely to get shingles (if you haven't had chickenpox, the document says, you won't get shingles). But there are also possible side effects from taking the vaccine, whose risk can be clearly quantified, and those are risks for everyone who takes the vaccine.

What seems to be a clear presentation of the likelihood of reducing a serious risk turns out to be so badly stated as to be almost worthless without more information.

Example 19 "Women are generally informed that mammography screening reduces the risk of dying from breast cancer by 25%. Does that mean that from 100 women who participate in screening, 25 lives will be saved? Although many people believe this to be the case, the conclusion is not justified. This figure means that from 1,000 women who participate in screening, 3 will die from breast cancer within 10 years, whereas from 1,000 women who do not participate, 4 will die. The difference between 4 and 3 is the 25% 'relative risk reduction.' Expressed as an 'absolute risk reduction,' however, this means that the benefit is 1 in 1,000, that is, 0.1%. Cancer organizations and health departments continue to inform women of the relative risk reduction, which gives a higher number—25% as compared to 0.1%—and makes the benefit of screening appear larger than if it were represented in absolute risks."

U. Hoffrage and G. Gigerenzer, "How to improve the diagnostic inferences of medical experts" in *Experts in science and society*

Evaluating risk uses all our reasoning skills. It's an important part of making good decisions.

24 Making Decisions

The skills you've learned here can help you make better decisions.

Making a decision is making a choice. You have options. Make a list for and against each choice—all the pros and cons you can think of. Make the best argument for each side. Then your decision should be easy: choose the option for which there is the best argument. Making decisions is no more than being very careful in constructing arguments for your choices.

But there may be more than two choices. Your first step should be to list all the options and give an argument that these really are the only options and not a false dilemma. Remember, too, that doing nothing is a choice, which may have serious consequences.

Suppose you do all that and you still feel there's something wrong. You see that the best argument is for the option you feel isn't right. You have a gut reaction that it's the wrong decision. Then you're missing something. Don't be irrational. You know when confronted with an argument that appears good yet whose conclusion seems false, you should show that the argument is weak or a premise is implausible. Go back to your pro and con lists. Get more information if you can.

Now that your reasoning has been sharpened, you can understand more, you can avoid being duped. And, we hope, you will reason well with those you love and work with and need to convince, and you will make better decisions. But whether you will do so depends not just on method, not just on the tools of reasoning, but on your goals, your ends. And that depends on virtue.

Writing Well

First comes clear thinking. Then comes clear writing.

You know how to evaluate claims, arguments, causal reasoning, and explanations. When you want to present your own ideas, use these skills to evaluate your own writing.

Writing well requires practice, unlearning many of the tricks of padding out essays. You'll have lots of opportunities in your studies and your work. Here's a summary of some points that will be helpful.

- If you don't have an argument, literary style won't salvage your essay.

- If the issue is vague, use definitions or rewrite the issue in order to present a precise claim to deliberate.

- Don't make a clear issue vague by appealing to some common but meaningless phrase, such as "This is a free country."

- Slanters turn off those you might want to convince; fallacies just convince the careful reader that you're dumb or devious.

- Beware of questions used as claims. The reader might not answer them the way you do.

- Your argument won't get any better by weaseling with "I believe that" or "I feel that." Your reader probably won't care about your feelings, and your feelings won't establish the truth of the conclusion.

- Your reader should be able to follow how your argument is put together. Indicator words are essential.

- Your premises must be highly plausible, and there must be glue, some claim that connects the premises to the conclusion.

- Don't claim more than you actually prove.

- There is often a trade-off: You can make your argument strong, but perhaps only at the expense of a rather dubious premise. Or you can make all your premises clearly true, but leave out the dubious premise that is needed to make the argument strong. Given the choice, opt for making the argument strong. If it's weak, no one should accept the conclusion. And if it's weak because of an unstated premise, it's better to have that premise stated explicitly so it can be the object of debate.

- In making an argument, you might think you have a great one. All the premises seem obvious and they lead to the conclusion. But if you imagine someone objecting, you can see how to give better support for doubtful premises or make it clearer that the argument is valid or strong. Answering counterarguments in your writing also allows the reader to see that you haven't missed some obvious objections. Just make a list of the pros and cons. Then answer the other side. Your argument must be impervious to the questions "So?" and "Why?".

- Sometimes the best argument may be one which concludes that we should suspend judgment.

Use the critical abilities you've developed to read your own work. Learn to stand outside your work and judge it, as you would judge an argument made by someone else.

If you reason calmly and well, you will earn the respect of others, and you may learn that others merit your respect, too.

Index

Richard L. Epstein received his B.A. summa cum laude from the University of Pennsylvania and his Ph.D. from the University of California, Berkeley. He was a Fulbright Fellow to Brazil and a National Academy of Sciences scholar to Poland. He is the author of the textbook *Critical Thinking* as well as *Propositional Logics* (Kluwer), and *Predicate Logic* (Oxford) among many other books. Currently he is the head of the Advanced Reasoning Forum in Socorro, New Mexico.

Other books by Richard L. Epstein published by ARF

The Fundamentals of Argument Analysis

Cause and Effect, Conditionals, Explanations

Prescriptive Reasoning

Reasoning in Science and Mathematics

Reasoning and Formal Logic

The Advanced Reasoning Forum is dedicated to advancing the study and teaching of critical thinking, formal logic, philosophy of language, linguistics, and ethics.

www.AdvancedReasoningForum.org